Infectious Agents and Cancer

Anton G. Kutikhin • Arseniy E. Yuzhalin
Elena B. Brusina

Infectious Agents and Cancer

Springer

Anton G. Kutikhin
Kemerovo State Medical Academy
Kemerovo
Russia

Arseniy E. Yuzhalin
Kemerovo State Medical Academy
Kemerovo
Russia

Elena B. Brusina
Kemerovo State Medical Academy
Kemerovo
Russia

ISBN 978-94-017-8073-5 ISBN 978-94-007-5955-8 (eBook)
DOI 10.1007/978-94-007-5955-8
Springer Dordrecht Heidelberg New York London

Printed on acid-free paper

Springer is part of Springer Science+Business Media (www.springer.com)

I dedicate this book to my beloved Carmen, who is my one and only, and who is the only one who knows the true reason why I have chosen the path of cancer biologist. From the very beginning to the finish of this book, she was with me, supporting me incessantly and granting me the power to continue writing. It could not have been possible to create this book without you, my one love.

I also want to thank Evie, Leticia and Phoebe, who saved me in the hardest time of my life and who always told me I am able to go on in spite of any circumstances. This book keeps also a part of each of you.

Anton G. Kutikhin, Kemerovo

Preface

This book is devoted to the role of infectious agents in the etiology of human cancer. Over the years of cancer investigation, a lot of discoveries in this field were made, and many associations between various biological carcinogens and cancer risk were revealed. Some of them are credibly determined, thus these infectious agents (human papilloma virus, hepatitis B virus, hepatitis C virus, Epstein-Barr virus, human herpes virus 8, human T-cell lymphotropic virus 1, human immunodeficiency virus, Merkel cell polyomavirus, *Helicobacter pylori*, *Opisthorchis viverrini*, *Clonorchis sinensis*, *Schistosoma haematobium*) are recognized as carcinogens and probable carcinogens by International Agency for Research on Cancer (IARC). They are not considered in this book since they were perfectly reviewed in a brilliant monograph *Infections Causing Human Cancer* authored by Nobel Prize Laureate Harald zur Hausen. The aim of this monograph is to analyze associations of other infectious agents with cancer risk (thus, it is admissible to call them "unconventional" from the point of view of cancer biologist and epidemiologist). To the best of our knowledge, this is the first book devoted to this problem and highlighting its importance and topicality. Of note, virology is not considered in our monograph: although there are some viruses that can be connected with cancer but are not included into the IARC list (John Cunningham virus, herpes simplex virus-1 and -2, human cytomegalovirus, simian virus 40, xenotropic murine leukemia virus-related virus), we decided to leave them for the virologists and to concentrate our efforts on other infectious agents (bacteria, protozoa, helminths and fungi) since it will better correspond to the conception of this book. Firstly, we note the criteria of inclusion of infectious agents into the list of possible biological carcinogens, and then we briefly describe the general mechanisms of biological carcinogenesis related to the infectious agents analyzed in the monograph (again, we do not concern principles of viral carcinogenesis). In the second part of the book, we review and analyze the available published literature about the possibly carcinogenic infectious agents (firstly bacteria, then protozoa, and, at last, helminths). In the last chapters, we summarize the data about the role of the whole *systems of microorganisms* (oral and gut microbiota) in the local and distant carcinogenesis. Finally, a brief conclusion about the role of the "unconventional" carcinogenic infectious agents in cancer etiology, improving the conception of biological carcinogenesis is given. We prepared this book with the hope that it will be useful for the wide audience, particularly cancer researchers, epidemiologists, microbiologists, Ph.D., graduate and undergraduate students of biomedical faculties and their lecturers.

Department of Epidemiology
Kemerovo State Medical Academy
Kemerovo, Russian Federation

Anton G. Kutikhin

Department of Epidemiology, Research Institute for Complex Issues of Cardiovascular Diseases under the Siberian Branch of the Russian Academy of Medical Sciences, Kemerovo State Medical Academy, Kemerovo, Russian Federation

Arseniy E. Yuzhalin

Head of the Department of Epidemiology
Kemerovo State Medical Academy
Kemerovo, Russian Federation

Elena B. Brusina

Contents

Chapter 1
The Criteria of Inclusion of Infectious Agents in the List of Biological Carcinogens

Abstract In this chapter, the criteria of inclusion of infectious agents in the short and extended lists of biological carcinogens are discussed. We also propose the criteria for the credibility of the connection between the non-viral infectious agent and cancer.

Keywords Cancer • Infections • Infectious agents • Carcinogens • Biological carcinogenesis • International Agency for Research on Cancer (IARC) • Criteria of carcinogenicity • Classification • List • Association

To classify infectious agents not included in the list of the International Agency for Research on Cancer (IARC) according to their possibility to cause cancer, the following criteria may be proposed.

To be included into the short list of potential biological carcinogens both conditions described below should be fulfilled:

- Basic studies confirming the fundamental carcinogenic properties of the infectious agent (ability to cause mutations and epigenetic alterations, inhibit apoptosis, enhance growth, accelerate proliferation, support survival, inhibit differentiation, enhance angiogenesis, promote stroma restructuring and invasion, should be conducted (as on cell lines and as on animal models)
- The association between the infectious agent and cancer risk should be demonstrated in the large prospective well-designed epidemiological studies in various countries

To be included into the extended list of potential biological carcinogens one of conditions described below should be fulfilled:

- Basic studies confirming the fundamental carcinogenic properties of the infectious agent (ability to cause mutations and epigenetic alterations, inhibit apoptosis, enhance growth, accelerate proliferation, support survival, inhibit differentiation,

A.G. Kutikhin et al., *Infectious Agents and Cancer*, DOI 10.1007/978-94-007-5955-8_1,

enhance angiogenesis, promote stroma restructuring and invasion, should be conducted (as on cell lines and as on animal models)
- The association of the infectious agent with cancer risk should be shown in the epidemiological studies

In general, the criteria for the credibility of the connection between the non-viral infectious agent and cancer may be postulated as follows:

- It should be detected in significant share of tumor cases.
- It must be localized in malignant tissues significantly more frequently and/or in higher concentration compared to corresponding healthy tissues.
- It must colonize the tissue before cancer development, at every stage of multistep carcinogenesis or even when tissues are absolutely normal.
- It must have eminent carcinogenic activity on the molecular and cellular level (it must have an ability to cause mutations and epigenetic alterations, inhibit apoptosis, enhance growth, accelerate proliferation, support survival, inhibit differentiation, enhance angiogenesis, promote stroma restructuring and invasion).
- It must induce carcinogenesis on animal models, and it should be proved by different research groups.
- Etiotropic therapy should lead to the regression of tumor associated with infectious agent.
- The association between the infectious agent and cancer risk must be confirmed in large prospective well-designed epidemiological studies in various countries and desirably by different research groups.

Chapter 2
General Mechanisms of Biological Carcinogenesis

Abstract In this chapter, we briefly describe the general mechanisms of biological carcinogenesis. They can be divided into the three groups: metabolic mechanisms, immune mechanisms and mechanisms mediated by the carcinogenic activity of the bacterial and protozoan toxins. For the simplicity, we note only the very basic moments, since, according to our united opinion, only these issues are important for the conceptual understanding of the problem.

Keywords Cancer • Infections • Infectious agents • Carcinogens • Biological carcinogenesis • Mechanisms of carcinogenesis • Metabolism • Immunity • Immune system • Toxins

2.1 Metabolic Mechanisms

The action of these mechanisms is based on the conception that biological agents participate in the degradation, modification, biotransformation and detoxification of chemical compounds coming into the gastrointestinal tract after the consumption of food and drink ultimately metabolizing them to carcinogens. Infection by certain agents affecting the liver (for instance, *Helicobacter hepaticus* or *Fasciola hepatica*) may also have influence on the hepatic expression and activity of enzymes involved in carcinogen bioactivation (phase I enzymes) and detoxification (phase II enzymes) that can also modulate cancer risk. A number of members of oral and gut microbiota possess these properties. This is described in details in the corresponding sections. In addition, human gastrointestinal microbiota may also affect weight gain and promote obesity thus increasing the risk of obesity-associated malignancies.

A.G. Kutikhin et al., *Infectious Agents and Cancer*, DOI 10.1007/978-94-007-5955-8_2,

2.2 Immune Mechanisms

Chronic inflammation substantially increases the probability of neoplastic transformation of the surrounding cells, inducing mutations and epigenetic alterations by the activity of inflammatory molecules (interleukins, leukotriens, thromboxans, TNF-α, prostaglandins, etc.) through the formation of free radicals and DNA damage, and also leading to apoptosis inhibition, prevention of cell differentiation, stimulation of the proliferation, growth, survival, stroma restructuring, angiogenesis and invasion. Since infectious agents persisting in the organism may cause chronic inflammation, they can also promote local carcinogenesis. In addition, infection may promote growth and proliferation of cells without the functioning of the immune system (for instance, in the experiments with *Helicobacter bilis* and certain epithelial cell lines). Chronic inflammation can also specifically affect the functioning of the organ, for instance, promoting cholelithiasis and urolithiasis that increase the time of exposure of the gallbladder, bile ducts, urinary bladder and ureters to chemical carcinogens and carcinogenic bacteria. Antigen-stimulated lymphoproliferation that can have place in the case with *Borrelia burgdorferi* and *Chlamydia psittaci* may be classified as one more type of immune mechanisms as well. This type of mechanisms is possessed by almost all of the biological carcinogens. The parasite-induced modulation of the host immune response may also be associated with loss of regulatory mechanisms involved in the immunological surveillance against cancer, elevating risk of malignant tumor development.

2.3 Mechanisms Mediated by the Carcinogenic Activity of the Bacterial and Protozoan Toxins

In addition to the metabolic and immune mechanisms, a number of bacteria (enterotoxigenic *Bacteroidesfragilis*, certain *Escherichia coli* strains, *Helicobacter bilis*) and protozoa (*Toxoplasma gondii*) produced or contained in their cell wall their own toxins (fragilysin, matrilysin, cytolethal distending toxin, cycle inhibiting factor, cytotoxic necrotizing factor 1, lipopolysaccharide) possessing a carcinogenic activity, affecting cell-cell interactions, intracellular signal transduction or induction of mutations and epigenetic alterations that can influence vital cell processes (apoptosis, proliferation, survival, growth, differentiation, invasion). Intracellular protozoan (*Toxoplasma gondii*) may induce resistance to multiple mechanisms of apoptosis (Fas-dependent and Fas-independent CTL-mediated cytotoxicity, IL-2 deprivation, gamma irradiation, UV irradiation, calcium ionophores). So, bacterial and protozoantoxins may function like initiating or like promoting agents.

Chapter 3
The Role of Bacteria in Cancer Development

Abstract Nowadays, a plethora of information about the role of bacteria in cancer development is available. However, there is only one bacterium, *Helicobacter pylori*, which is recognized by IARC as an established human carcinogen. In this chapter, we analyze the information about other infectious agents which can cause cancer but are not included into the IARC list. For the convenience of the readers, we describe the role of bacteria in cancer of various systems of human body separately. Firstly, we analyze the role of *Helicobacter bilis*, *Helicobacter hepaticus*, *Campylobacter jejuni*, *Clostridium* spp., *Bacteroides* spp., *Salmonella typhi*, and *Streptococcus* spp. in cancer of the digestive tract, than we focus on the role of *Mycobacterium tuberculosis* and *Chlamydia pneumoniae* in cancer of the respiratory tract, continue the story with the role of various agents of genitourinary infections in cancer of the genitourinary tract, consider *Borrelia* spp. and *Chlamydia psittaci* as the potential causes of cancer of the lymphopoietic tract, and, finally, describe the bacteria which are only anecdotally associated with malignant tumors.

Keywords Cancer • Infections • Infectious agents • Carcinogens • Bacteria • Helicobacter • Streptococcus • Chlamydia • Salmonella • Borrelia

3.1 The Role of Bacteria in Cancer of the Digestive Tract

3.1.1 Helicobacter bilis

Helicobacter bilis (*H. bilis*) is a gram negative, nonsporulating, microaerophilic bacteria with fusiform to slightly spiral form, measuring 0.5–5 μm with 3–14 multiple bipolar sheathed flagella and periplasmic fibers wrapped around the cell. It was isolated from the hepatic, extrahepatic and intrahepatic bile and tissue. It is known that *H. bilis* colonizes distal small intestine, caecum and large intestine, and the

A.G. Kutikhin et al., *Infectious Agents and Cancer*, DOI 10.1007/978-94-007-5955-8_3,

hepatic and extrahepatic infection occurs through enterohepatic route. *H. bilis* was not detected in the stomach and is related to so-called "nongastric *Helicobacter*". *H. bilis* virulence factors include colonization, persistence and disease inducing factors.

The first study devoted to the possible role of *H. bilis* in cancer etiology was published in 1998, when Fox et al. (1998) identified several strains of this bacterium in bile samples and resected gallbladder tissues by PCR analysis (recovery of Helicobacter sp. from frozen specimens was unsuccessful) from 22 out of 46 Chilean patients with chronic cholecystitis. The idea that *H. bilis* may be involved in the cancer development postulated in this research was continued in the investigation of Matsukura et al. (2002) who collected and analyzed the PCR bile samples from 45 Japanese and 40 Thai patients divided into the three groups: 15 Japanese and 14 Thai subjects suffered from bile duct or gallbladder cancer, 16 Japanese and 26 Thai subjects with gallstones and/or chronic cholecystitis, and 14 Japanese subjects in whom non-biliary diseases were diagnosed. The results revealed that 24 out of 29 indicated cancer cases, 18 out of 36 gallstones/chronic cholecystitis cases and 4 out of 14 patients with non-biliary diseases were positive for *H. bilis*. The authors of the study calculated the odds ratio and found that among Japanese subjects, the risk of *H. bilis* identification in bile duct of gallbladder cancer cases was 6.5 times higher (95% CI=1.09–38.63) than in patients with gallstones/chronic cholecystitis. Concerning the Thai population, the odds ratio for *H. bilis* identification in the bile of bile duct/gallbladder cancer cases was 5.86 (95% CI=1.31–26.33) compared to the bile samples of the subjects with gallstones/chronic cholecystitis. In addition, the share of *H. bilis*-positive individuals among patients with gallstones/chronic cholecystitis in these two studies was remarkably similar (about 50%). Concerning bile duct/gallbladder cancer, this share was significantly higher (about 83%). In 2004 in Japan, Murata et al. (2004) tried to replicate the results of the previous two studies, performing PCR analysis of the DNA extracted from 34 archival paraffin-embedded gallbladder specimens, which were obtained from 14 patients with bile duct/gallbladder cancer, 16 subjects with cholecystolithiasis and 4 patients suffering from pancreatic cancer. The percentage of *H. bilis*-positive cases was lower than in investigations of Fox et al. (1998) and Matsukura et al. (2002) reaching 29% (4 out of 14) in cancer patients and 12.5% (2 out of 16) in subjects with cholecystolithiasis. Nilsson et al. (2006) found DNA of *H. bilis* using PCR in 60% of the Helicobacter-positive gastroduodenal tissue samples of the Swedish pancreatic cancer patients (one-third of them were Helicobacter-positive in gastroduodenal specimens). Finally, the last case-control study devoted to this problem was carried out by Pandey et al. (2010). In India, this research group recruited the study sample of 54 gallbladder cancer cases and 55 controls with cholelithiasis and identified *H. bilis* by PCR in tissue samples of 32 out of 54 cases (59%) and 32 out of 55 controls (58%), suggesting that this bacterium does not play a substantial role in the etiology of gallbladder cancer. They also performed a meta-analysis of the results of three studies published earlier on *H. bilis* and gallbladder cancer. The final odds ratio was 1.92 (95% CI 1.06–3.47). Heterogeneity among the studies was relatively low ($\chi^2=8.69$, $p=0.03$). In total, 73 out of 124 cases and 38 out of 87 controls were positive

Table 3.1 The connection between *H. bilis* and bile duct/gallbladder cancer in epidemiological studies

Authors, year, population	Sample size	Methods of detection	Share of positive individuals, OR, 95% CI
Matsukura et al. (2002), Japanese, Thai population	15 Japanese and 14 Thai cancer cases, 16 Japanese and 26 Thai controls with gallstones and/or chronic cholecystitis, and 14 Japanese controls with non-biliary diseases	PCR	83% (cases), 50% (controls with gallstones/chronic cholecystitis), 28.5% (other controls) Among Japanese subjects: 6.5 (1.09–38.63) Among Thai subjects: 5.86 (1.31–26.33)
Murata et al. (2004), Japanese population	14 cancer cases, 16 controls with cholecystolithiasis	PCR	29% (cases), 12.5% (controls)
Pandey et al. (2010), Indian population	54 cancer cases, 55 cholelithiasis controls	PCR	59% (cases), 58% (controls) In meta-analysis of all three studies described in the table: 59% (cases), 43.5% (controls) 1.92 (1.06–3.47)

for *H. bilis*. Unfortunately, the results of this meta-analysis can be imprecise because the authors mixed the control groups which consisted of patients with non-biliary diseases and with chronic cholecystolithiasis. It is also important to note that the methods of *H. bilis* detection used in these investigations were different (PCR in bile samples and in paraffin-embedded gallbladder tissues has distinct sensitivity). In addition, some differences between populations (Japanese, Thai and Indian) may play a role (Table 3.1).

In 2005, Yanagisawa et al. (2005) conducted the first basic research devoted to the carcinogenic properties of *H. bilis*. They found that *H. bilis* strains increased the expression of MMP-7 on colon epithelial cells HT29 compared to uninfected control cells, and this effect was bacterial dose-dependent. Since MMP-7 facilitates bacterial invasion, tumor invasion and metastasis by degrading the extracellular matrix, removing the physical barriers and activating the residing proteins, the authors suggested that this increase of its expression may be a probable mechanism of *H. bilis* carcinogenicity. Maggio-Price et al. (2002, 2005, 2006) discovered that *H. bilis* infection induces severe intestinal inflammation, hyperplasia, dysplasia and colitis development in *mdr1a* –/– mice and in *Smad3* –/– mice (in latter case, mucinous adenocarcinomas develop on two-thirds of cases). *H. bilis*-induced dysplastic changes in colonic epithelial cells of *mdr1a* –/– mice and *Smad3* –/– mice correlated with increased proliferation of epithelial cells, T-helpers, macrophages, increased expression of the MHC class II, Cox-2, CD4, F4/80, proinflammatory cytokines IL-1β, IL-1α, IL-6, IFN-γ, TNF-α, and also with increased expression of the *c-myc* oncogene. Takayama et al. (2010) prepared HuCCT-1, the human bile duct cancer cell line, cocultured it with *H. bilis*, and then found that NF-κB, E2 transcription factor (E2F), cyclic AMP response element (CRE) activity, production of vascular endothelial growth factor (VEGF) and angiogenesis in *H. bilis*-infected HuCCT-1 cell line were enhanced compared to control HuCCT-1 cell line. As pathways depending on the functioning of the indicated proteins are the key for the carcinogenesis, the authors proposed that *H. bilis*-infection may play an important role in the development of malignancies in the biliary tract.

Futhermore, the association of cholecystitis/cholelithiasis with bile duct/gallbladder cancer was demonstrated by Andreotti et al. (2011, 647 cancer cases and 959 healthy subjects, OR=1.7, 95% CI=1.1–2.9), Hsing et al. (2007, 627 cancer cases, 1,037 gallstones cases, 959 healthy controls, OR=23.8, 95% CI=17.0–33.4 and 8.0, 95% CI=5.6–11.4 for gallbladder and bile duct cancer, respectively), Lowenfels et al. (1985, 131 gallbladder cancer cases, 2,399 controls, OR=4.4, 95% CI=2.6–7.3 for non-Indian population and RR=20.9, 95% CI=8.1–54), Sheth et al. (2000), and in an extensive study by Bansal and Sonnenberg (1996). In addition, the connection of these pathologies with pancreatic cancer was detected by Bansal and Sonnenberg (1996), with colorectal cancer was noted in the meta-analysis of Xu et al. (2009, OR=2.68, 95% CI=1.93–3.72), with rectal cancer in the meta-analysis of Chiong et al. (2012, OR=1.33, 95% CI=1.02–1.73), and with the small intestinal cancer, prostate cancer and kidney cancer in the meta-analysis of Tavani et al. (2012, OR=3.96, 1.36 and 1.57, respectively). Although there are a number of other population and animal studies devoted to the correlation of chronic inflammatory conditions of the gallbladder and bile ducts with cancer of these sites, the positive association is clear in the conception of chronic inflammation as one of the main cancer promoters.

Obviously, there is an open ended question on the influence of *H. bilis* on the gallbladder cancer development. On the one hand, the results of the basic studies conducted on the cell lines and population studies give the possibility to suggest that the true relation can exist. On the other hand, the results of the latter study and less extensive investigations carried out create some doubts in the validity of the proposed theory. Future studies with the appropriate design can solve this problem. Firstly, the study sample should be divided into the three groups: Bile duct/gallbladder cancer cases, patients with cholelithiasis/chronic cholecystitis and healthy controls or subjects with non-biliary diseases. This stratification will make it feasible to analyze whether *H. bilis* has the "true oncogenic role" releasing substances that affect the functioning of oncogenes or tumor suppressor genes, or whether it has only pro-inflammatory role and increase cancer risk only through inflammatory mechanisms, that are the same that provoke gallstones formation and cholecystitis development. Secondly, PCR of the DNA extracted from the bile and gallbladder tissues should be used instead of culturing methods. PCR of the DNA obtained from the bile samples can be significantly more sensitive than PCR of DNA collected from the gallbladder tissues. Thirdly, further basic studies on the cell lines and animal models should be performed to discover new carcinogenic properties of *H. bilis*. At the moment, it is possible to suggest that this bacterium may cause bile duct/

gallbladder cancer, but there are not enough arguments to include it in the short roster of the infectious agents that may be a reason of cancer development but are not included in the International Agency for Research on Cancer (IARC) list. However, *H. bilis* may certainly be included in the extended roster.

3.1.2 *Helicobacter hepaticus*

Helicobacter hepaticus (*H. hepaticus*) is motile, gram negative, microaerophilic bacterium, curved or spiral in shape, 0.2–0.3 μm in diameter, 1.5–5 μm in length. It has bipolar sheathed flagella (one at each end) but lacks the periplasmic fibers. *H. bilis*, *H. hepaticus* colonizes the enterohepatic tissues. In 1994, Ward et al. (1994a, b) and Fox et al. (1994, 1995) reported the isolation of a new species of *Helicobacter* (provisionally designated *Helicobacter hepaticus* sp. nov.) that selectively and persistently colonized the hepatic bile canaliculi of mice, the intrahepatic biliary system and large bowel, causing a morphologically distinctive pattern of chronic, active hepatitis characterized by hepatocyte loss, mild cirrhosis, cholangitis, hepatocytomegaly, cholangiocellular hyperplasia, persistent hepatocyte proliferation, and associated with a high incidence of preneoplastic foci followed by hepatocellular adenoma and carcinoma. They created the first animal model of the *H. hepaticus* infection and concluded that *Helicobacter hepaticus* can be a cause of rodent hepatocellular tumors. The features of the pathomorphological, immunological and clinical characteristics of the chronic proliferative hepatitis caused by persistent *H. hepaticus*-infection were described in more detail 2 years later in the comprehensive studies of Fox et al. (1996a, b). In the same year, Ward et al. (1996a) discovered an *H. hepaticus* protein that reacted with heat-shock protein 70 (HSP70) antibodies. The authors also noted the increased expression of HSP70 RNA and protein in hepatocytes, and reported a presence of serum antibodies to HSP70 in infected mice along with hepatic infiltration of CD3+ (T) lymphocytes, underlining the possible autoimmune component in the development of chronic active hepatitis and liver tumors caused by this bacterium. In addition, a number of research groups including the groups of Ward et al. (1996b), Foltz et al. (1998), Li et al. (1998), Kullberg et al. (1998), and others, discovered a potential role of *H. hepaticus* in the development of inflammatory bowel diseases including typhlitis, colitis, and proctitis in various mouse lines.

Chomarat et al. (1997) investigated the impact of *H. hepaticus*-infection on the hepatic expression and activity of enzymes involved in carcinogen bioactivation (phase I enzymes) and detoxification (phase II enzymes). They found no significant difference in total cytochrome P450 (CYP) content between infected and control mice, but noted the increase in CYP2A5 and CYP1A2 expression in mice aged over 12 months. In addition, hepatocytes often co-expressed CYP2A5 and formazan deposition. The infected mice were characterized by higher glutathione S-transferase (GST) activity, but in contrast, catalase and glutathioneperoxidase activities and reduced glutathione content were decreased in 3–9 months of disease in infected mice compared to controls. The authors concluded that alterations in the expression of the enzymes responsible for xenobiotic transformation and antioxidant defense may contribute to the *H. hepaticus*-caused tumor formation via enhanced production of mutagenic reactive oxygen species (ROS). Diwan et al. (1997) proved the previous results of carcinogenicity of *H. hepaticus* and found that chronic *H. hepaticus* infection significantly stimulated cyclin D expression, suggesting its role in inflammation, abnormal cell growth, and early neoplasia. Sipowicz et al. (1997) noted the lack of *p53* and *ras* mutations in Helicobacter hepaticus-induced liver tumors in A/JCr mice and postulated that other genes may be targeted or *H. hepaticus*-infection causes liver cancer through pathways other than direct damage to DNA. Hailey et al. (1998) found that the oncogenic potential of *H. hepaticus* is similar in B6C3F1 mice as in other mice investigated previously. Ramljak et al. (2000) revealed that epidermal growth factor (EGF) and transforming growth factor-α (TGF-α) were increased at the earliest stage of infection, with a further raise in EGF in tumors. Cyclin D1, cyclin-dependent kinase 4 (CDK4), and c-Myc oncogenic protein were significantly elevated in all infected livers, and in tumors they were even greater. An increase in cyclin D1/CDK4 complex was also demonstrated in neoplasms, and its functionality was confirmed by an elevation in the hypophosphorylated Rb protein ratio. Diwan et al. (2000) observed the increase of Grb2 protein in infected livers as well as in preneoplastic foci, adenomas, and carcinomas in comparison with control livers. In addition, infected livers were characterized by the elevation of total Ras protein. The authors hypothesized that upregulation of Grb2 is an early event in liver carcinogenesis, and further upregulation of Ras can facilitate transmission of amplified signal from growth factors via Grb2 protein. Dieleman et al. (2000) revealed that IL-10–/– mice monoassociated with *H. hepaticus* for up to 16 weeks did not have the colitis or increased IL-12 production. In addition, IL-10-knockout mice had no significant difference in weight loss, mortality rate, histologic scores, colonic IL-12 secretion, or T-cell activation with or without *H. hepaticus*, so *H. hepaticus* did not induce or potentiate disease in IL-10–/– mice.

Erdman et al. (2003a, b) demonstrated that *H. hepaticus*-infected mice without recombinase-activating gene 2 (Rag2–/–) mice were characterized by rapidly developing colitis and large bowel carcinoma, demonstrating a link between microbial inflammation and cancer in the lower bowel. Based on the results of this study, this research group also suggested that innate immune dysregulation may have a key role in inflammatory bowel disease and its progression to cancer. *H. hepaticus*-infected wild-type mice did not develop inflammation or carcinoma showing that lymphocytes are mandatory to preventing bacterially induced cancer at this site. Adoptive transfer with $CD4^{+}CD45RB^{lo}$ $CD25^{+}$ regulatory T cells into Rag-deficient hosts significantly inhibited *H. hepaticus*-induced inflammation and the development of cancer, suggesting that the ability of $CD4^{+}$T cells to protect against intestinal cancer correlates with their ability to reduce bacterially induced inflammatory bowel disease. In the same year, Erdman et al. (2003a, b) infected 129/SvEv Rag2–/– mice intragastrically with *H. hepaticus*, and then treated them with adoptive transfer of CD4+CD45RBloCD25+-regulatory T cells from 1, 3, and 12 months after the infection. Mice in which the regulatory cells were transferred from 4 or 12 weeks after H. hepaticus infection were characterized by less severe inflammatory bowel disease and significantly lower risk of colon cancer during the 8-month observation period, compared with infected mice that had not received those cells. In addition, transfer of regulatory T cells into chronically infected mice with established cancer reduced the severity of colitis, epithelial dysplasia, and cancer, but did not eliminate all the tumors. Regulatory cells lacking anti-inflammatory cytokine IL-10 were unable to inhibit inflammatory bowel disease, dysplasia, or cancer, suggesting that IL-10 was required for the protective effects of indicated lymphocytes.

Avenaud et al. (2003), based on their monitoring of 40 *H. hepaticus*-infected and 40 control A/J mice, proposed that bacterial strain or environmental conditions may be a key determinant of the progression of hepatocarcinogenesis. Rogers et al. (2004) described the course and pathogenesis of premalignant disease in A/JCr mice infected with *H. hepaticus* and revealed that mice which were infected at or before 3 weeks of age, but not at 12 weeks, developed the disease. Male mice were most affected and suffered from lobular necrogranulomatous and chronic active hepatitis, while females usually developed chronic persistent hepatitis. It was concluded by the authors that gender and time of bacterial exposure were the key determinants of *H. hepaticus* disease outcomes. Maggio-Price et al. (2002) discovered that *H. hepaticus* delays the onset of spontaneous colitis and attenuates the disease in *mdr1a* –/– mice and proposed that *H. hepaticus* infection may prevent *H. bilis*-inducedcolitis. However 3 years later, Maggio-Price et al. (2005) unexpectedly demonstrated that dual-infected mice are characterized by a higher frequency of high-grade crypt dysplasia and invasive adenocarcinoma, and suggested the possible mechanism for this occurrence. According to their hypothesis, *H. hepaticus*, in delaying the development of colitis, extends the amount of time for the transformation of epithelial cells. They noted that there was an increased expression of c-myc and interleukin-1α/β in colonic epithelial cells of dual-infected animals compared to uninfected wild-type and *mdr1a* –/– mice.

Yanagisawa et al. (2005) revealed that as in the case with *H. bilis*, *H. hepaticus*-infection elevated expression of MMP-7 on colon epithelial cells HT29 compared to uninfected control cells, and this effect was bacterial dose-dependent. Rao et al. (2006) found that Rag2-deficient C57BL/6 ApcMin/+ mice infected with *H. hepaticus* were characterized by (1) development of breast cancer in female animals and (2)enhanced intestinal adenoma multiplicity by a TNF-α-dependent mechanism. The mammary and intestinal tumor development and the increase in proinflammatory weresuppressedbyadoptivetransferofIL-10-competentCD4$^+$CD45RB^{lo}CD25$^+$regulatory T cells, and prior exposure of donor mice to *H. hepaticus* significantly elevated the antitumor potency of these cells. Besides, the indicated IL-10-competent CD4+CD45RBloCD25+regulatory T cells from the *H. hepaticus*-exposed donor mice, that was transferred into the recipient mice, suppressed tumorigenesis more effectively than in donor mice, irrespective of the tumor etiology. Therefore the research group proposed that infections with enteric pathogens such as *H. hepaticus* may increase the potency of regulatory T cells protecting against epithelial cancers later in life. In addition, the authors demonstrated the possibility that dysregulated gut microbial infections in humans may lead to cancer in anatomically distant organs such as the human breast. Ge et al. (2007) infected A/JCr mice with an isogenic mutant of *H. hepaticus* lacking cytolethal distending toxin (CDT) activity and revealed that the CDT mutant-infected mice did not have hepatic dysplastic nodules at 10 months after inoculation in contrast to those that were infected by wild-type *H. hepaticus*. At 4 months after inoculation, the mice which were infected with wild-type *H. hepaticus* were characterized by increased hepatic transcription of proinflammatory *TNF*-a, *IFN-γ*, *Cox-2*, growth mediators *IL-6* and *TGF*-α, anti-apoptotic proteins *Bcl-2* and *Bcl-XL*, and by enhanced hepatocyte proliferation compared to the control or the CDT mutant-infected mice. Furthermore, male mice that were infected with wild-type *H. hepaticus* had upregulated hepatic mRNA levels of RelA (p65), p50, *GADD45*b and *c-IAP1* (components of the NF-kB pathway) in comparison with the CDT mutant-infected mice. At 10 months after inoculation, infection with wild-type *H. hepaticus* was associated with significant upregulation of *IL-6* mRNA. Liyanage et al. (2010) demonstrated that H2AX was phosphorylated, the expression of pro-apoptotic Bax protein was upregulated and Bcl-2 expression was decreased in the 24 h postexposure of INT407 cells to CDT of *H. hepaticus*. In addition, cytochrome c was released from mitochondria after 12–24 h of exposure, and caspases 3/7 and 9 were activated, that confirmed the role of CDT in mediation of the apoptosis in specifically treated cells.

Diwan et al. (2008) compared the intragastric and intraperitoneal ways in which the *H. hepaticus* infection spread in different mice strains, and revealed that the intraperitoneal way is significantly associated with the increased number of liver tumors including carcinomas. In addition, the intragastric way of weanling mice resulted in low incidence of hepatitis and lower count of liver tumors and this suggested that high incidence of liver tumors in mice infected with *H. hepaticus* requires perinatal exposure. García et al. (2008) used F1 hybrid mice derived from A/J and C57BL/6 matings to investigate their susceptibility and hepatic gene expression profile in response to *H. hepaticus* infection. *H. hepaticus*-infected parental strains including A/J and C57BL/6 mice, as well as F1 mice, developed significant hepatitis. Hepatocellular carcinomas or dysplastic liver lesions were observed in 69% of *H. hepaticus*-infected F1 male mice, and the bacterium was isolated from the hepatic tissues of all F1 mice with liver tumors. They found that the liver tumors with severe hepatic steatosis developed in the livers with high hepatitis scores. Fatty acid synthase and steaoryl-coenzyme A desaturase, the two rate-limiting enzymes in lipogenesis, were upregulated in neoplastic relative to dysplastic livers, suggesting a synergistic interaction between hepatic steatosis and infectious hepatitis leading to hepatocellular carcinoma. In addition, the authors proposed that immune response genes such as major histocompatibility genes *H2-Aa*, *H2-Eb1*, *H2-Ab1*, and other genes included the *Ikbkg*, *Stat1*, *Tgfb1*, *Ccl5*, *Cd44*, *Pla2g7*, *Cyba*, and *Rac2*.

Nagamine et al. (2008a) created another *H. hepaticus* model of colon tumorigenesis. This research group reported that BALB-RagMin mice had an increase in tumors in the small intestine relative to their BALB-Min counterparts, and proposed that the adaptive immune system plays a major role in creating a resistance to cancer of the small intestine in this model. *H. hepaticus*-infected BALB-RagMin mice developed moderate hepatitis, moderate typhlitis, and mild colitis, and these mice had a 14-fold elevation in colon tumor incidence relative to uninfected BALB-RagMin mice. In another study published in the same year, Nagamine et al. (2008b) investigated the interaction between carcinogenic effects of azoxymethane (AOM) and *H. hepaticus*, and found that infection with *H. hepaticus* prior to AOM-treatment did not result in colitis-associated adenocarcinomas but led to a significant increase in the incidence of AOM-induced colon tumors relative to AOM treatment alone, suggesting that *H. hepaticus* promotes AOM-induced tumorigenesis in the BALB/c-IL10 null mouse which were used in this research. Watanabe et al. (2008) demonstrated that single adoptive transfer of regulatory T cells prior to *H. hepaticus*-infection prevented the development of colitis despite persistent infection in recipient mice but, when infected mice were injected with anti-CTLA-4 antibody along with regulatory T cells transfer, they suffered from severe acute colitis and had increased expression levels of IFN-γ and IL-2. These results indicated that blockade of CTLA-4 by specific antibodies clearly abrogates the ability of the regulatory T cells to suppress innate immune-driven colitis.

Theve et al. (2008) investigated the influence of sex hormones on hepatitis in *H. hepaticus*-infected A/Jcr mice and found that castrated males and those receiving the competitive androgen receptor antagonist flutamide had significantly less severe disease. Transcriptional alterations involved both feminine (cytochrome P450 4a14) and masculine (cytochrome P450 4a12 and trefoil factor 3) genes, and also gender-neutral genes (H19 fetal liver mRNA, lipocalin 2, and ubiquitin D). Hepatitis was associated with increased unsaturated C18 long-chain fatty acids (oleic acid and linoleic acid) in relation to saturated stearic acid. Sterzenbach et al. (2008) were the first who investigated the role of *H. hepaticus* motility and flagella in intestinal colonization.

The flaA_1 flaA_2 double mutants and fliA mutants did not produce detectable amounts of flaA and possessed severely truncated flagella. Both mutants were non-motile and unable to colonize mice. Nevertheless, mutants that had either the flaA gene knockedout produced flagella morphologically similar to those of wild-type bacteria and expressed FlaA and FlaB. Of note, flaA_1 mutants which had flagella but were characterized by reduced motility did not colonize mice, indicating that motility is required for intestinal colonization by *H. hepaticus* and that the presence of flagella alone is not sufficient.

Erdman et al. (2009) showed that infection of Rag2−/− mice with Helicobacter hepaticus resulted in accumulation of macrophages and neutrophils in the colon, and, consequently, in upregulation of tissue inducible nitric oxide synthase (iNOS) expression at the site of infection and increased nitric oxide (NO) production. Progressive development of severe inflammation, hyperplasia, dysplasia, and cancer was in line with these changes. Administration of an iNOS inhibitor prevented NO production, diminished epithelial pathology and inhibited the onset of cancer. The presence of Gr-1^{+}neutrophils and high TNF-α expression in colon were required for increased iNOS expression and cancer, whereas IL-10 downregulated TNF-α and iNOS expression and suppressed cancer. Regulatory T cells also decreased iNOS expression and reduced cancer formation. Chow and Mazmanian (2010) revealed a protective role for the type VI secretion systems (T6SS) of *H. hepaticus*. T6SS mutants displayed increased intracellular numbers of bacteria within intestinal epithelial cells (IECs) and during intestinal colonization. In addition, $CD4^+$ T cells from T6SS mutant-colonized rodents produced increased IL-17 in response to IECs presenting *H. hepaticus* antigens. Co-culture experiments with IECs presenting *H. hepaticus* antigens showed higher levels of pro-inflammatory cytokines such as IL-17 when these IECs were incubated with T cells from T6SS mutant-colonized animals compared to wild-type-colonized animals. As a consequence, it was concluded by the authors that *H. hepaticus* interacts with IECs and uses the T6SS to set a balanced host relationship by limiting colonization and intestinal inflammation. Fox et al. (2010) demonstrated that intestinal colonisation by *H. hepaticus* is sufficient to promote aflatoxin- and HCV transgene-induced hepatocellular carcinoma (HCC). From its localization in the intestinal mucus layer, *H. hepaticus* activated NF-κB-regulated networks associated with innate and T helper 1 (Th1)-type adaptive immunity both in the lower bowel and liver. They also found indicators of tumor progression that included hepatocyte turnover, Wnt/β-catenin activation and oxidative injury with decreased phagocytic clearance of damaged cells.

Woodworth et al. (2010) tried to determine whether dietary fish oil (DFO) enriched with long-chain omega-3 fatty acid docosahexaenoic acid can reduce experimentally induced colitis and colon cancer risk in a SMAD3−/− mice exposed to *H. hepaticus*. However, contrary to their expectations, the DFO induced severe colitis and adenocarcinoma formation, and its consumption was associated with decreased $CD8^+$cell frequency and diminished CD69 expression on $CD4^+$ and $CD8^+$ T-cell populations. Mice consuming DFO also exhibited higher $FoxP3^+CD25^+CD4^+$ T regulatory cell frequency, FoxP3 expression, and altered L-selectin expression during infection. The authors concluded that DFO-fed mice may be less successful to develop an immune response to *H. hepaticus* infection, increasing colon cancer risk. García et al. (2011) showed that diethylnitrosamine (DEN)-treated *H. hepaticus*-infected mice with knockout gene of constitutive androstane receptor exhibited increased numbers of liver lobes with dysplasia and neoplasia and increased multiplicity of neoplasia in comparison with similarly treated wild-type mice. Enhanced tumor promotion was associated with decreased hepatic expression of P450 enzymes CYP2B10 and CYP3A11 and increased serum concentrations of chenodeoxycholic acid, suggesting that hepatocarcinogenesis may be enhanced by an impaired metabolic detoxification of endobiotics and a persistent microbial-induced immune response.

The first epidemiological study devoted to the problem was conducted in Nepal by Pradhan and Dali (2004). According to their data, *H. hepaticus*-infection was found in 82% of 100 gallbladders obtained from patients after cholecystectomy and in 85.7% of malignant cases (6 out of 7). Hamada et al. (2009) collected 126 bile samples from Japanese patients with cholelithiasis, cholecystitis, gallbladder polyps, and other nonbiliary diseases. As in the case with *H. bilis*, attempts to cultivate *H. hepaticus* were unsuccessful. However, *H. hepaticus* was detected in bile samples with nested PCR. These results were confirmed by in situ hybridization, and IgG antibodies to *H. hepaticus* in bile samples were identified by Western blotting. *H. hepaticus* was detected in 40 (32%) of the total 126 samples, and patients with cholelithiasis (41%) and cholecystitis with gastric cancer (36%) had significantly higher prevalence of *H. hepaticus* infection than samples from patients with other

Table 3.2 The connection of *H. hepaticus* with bile duct/gallbladder cancer cases in epidemiological studies

Authors, year, population	Sample size	Methods of detection	Share of positive individuals, OR, 95% CI
Pradhan and Dali (2004), Nepal population	7 cancer cases, 93 controls after the cholecystomy	Warthin Starry Silver staining	85% (cases), 82.5% (controls)
Hamada et al. (2009), Japanese population	28 cancer cases, 60 cholelithiasis controls, 38 controls with other diseases	PCR	36% (cases), 41% (controls with cholelithiasis), 13% (controls with other diseases)
Shimoyama et al. (2010), Japanese population	18 cancer cases, 55 controls with cholelithiasis, 34 asymptomatic controls	Western blotting	38.8% (cases), 18.2% (controls with cholelithiasis), 13.1% (asymptomatic controls)

diseases that is concordant with findings of Maurer et al. (2005) who found that monoinfections of C57L mice with *H. hepaticus* after 8 weeks of observance led to a cholesterol gallstones formation in 40% of cases. Finally, Shimoyama et al. (2010) collected serum samples from 55 Japanese patients with cholelithiasis, 18 subjects with bile duct or gallbladder cancer, 19 individuals with pancreatic cancer, 34 asymptomatic controls, and analyzed them by Western blot using a *H. hepaticus*-specific antigen. According to their data, prevalence of anti-*H. hepaticus*-antibodies in patients with bile tract cancer was 38.8% significantly higher than in control subjects (13.1%), in patients with cholelithiasis (18.2%) and with pancreatic cancer (10.5%) (Table 3.2).

To sum up, the role of *H. hepaticus* in carcinogenesis is rather interesting and equivocal. On the one side, its role in the development of colitis, benign and malignant tumors of the colon, as well as its significance in hepatitis and hepatocarcinogenesis, is obvious, and many discoveries in this field were carried out using the infected mice models. As written above, these models were extremely useful in investigating the role of apoptosis, oncogenes, immune response, endo- and xenobiotic transformation, antioxidant defense enzymes, sex hormones, and other fundamental processes in the etiopathogenesis of the indicated tumors. On the other side, the population studies that were conducted can provide only limited evidence that *H. hepaticus* may play a role in the bile duct/gallbladder cancer development. Certainly, further epidemiological studies are necessary to shed light on the problem of *H. hepaticus* hazard for humans. Investigations on cell lines in the future are also required to precisely determine the mechanisms of oncogenic action of *H. hepaticus*. The situation with *H. hepaticus* is similar with *H. bilis*: It is possible to suggest that this bacterium may cause bile duct/gallbladder cancer, but the reasons for including it in the short roster of the infectious agents as a cause of cancer development and not including it to the IARC list are insufficient. Nevertheless, *H. hepaticus* may certainly be included in the extended list.

3.1.3 *Campylobacter jejuni*

Campylobacter jejuni (*C. jejuni*) is small (1.5–6.0 μm long and 0.2–0.5 μm wide) Gram-negative spiral bacterium. It is one of the most important causes of bacterial diarrhea worldwide since it is contracted from untreated water, milk and meat. To date, there are only a few studies in which the role of *Campylobacter jejuni* (*C. jejuni*) in the etiology of malignancies is discussed. The first was the research of Lecuit et al. (2004) who obtained the frozen intestinal tissue from a patient with immunoproliferative small intestinal disease (IPSID) dramatically responsible for antibiotics (amoxicillin, metronidazole and clarithromycin) in combination with omeprazole, and identified the presence of *C. jejuni* using PCR, fluorescence in situ hybridization (FISH) and immunohistochemical assay of biopsy specimen. Thereafter, they conducted a follow-up retrospective analysis of archival intestinal biopsy specimens from six more patients with IPSID, and found *C. jejuni* by FISH and immunohistochemical assay in four out of six of them. Attempts of PCR detection were unsuccessful possibly because of fixation failure together with the age of the samples. The authors concluded in the results based on their findings that *C. jejuni* may be associated with the IPSID development. Beltinger et al. (2008) performed a model of initial C. jejuni pathogenic effects on intestinal epithelium. All the strains they used translocated across HCA-7 cell line monolayers but only a few of them invaded the cells, breaking monolayer resistance over a 6 h period. It was accompanied by elevated release of lactate dehydrogenase, increasing permeability to [3H] mannitol, enhanced IL-8 and prostaglandin E2 (PGE2) production, and ultrastructural disruption of tight junctions, with rounding and lifting of cells off the filter membrane. Finally, in 2010, a population study in Sweden was carried out by Brauner et al. (2010) where the study sample included all individuals in Stockholm County who tested positive for *C. jejuni* between 1989 and 2006. The cohort of 16,276 individuals generating a sum total of 124,387 years was followed-up until December 31, 2007 for the occurrence of cancer, and the background population was a control group. When the first year of follow-up was excluded, the overall risk of cancer was not higher in a *C. jejuni*-positive group during the 10 years, and there was no increased risk of cancer in the gastrointestinal tract, but the risk of melanomas and squamous cell skin cancer was significantly augmented [standard incidence ratio (SIR) = 1.84 (95% CI 1.27–2.57) and 1.52 (95% CI 1.01–2.19), respectively], but a significantly decreased risk of respiratory cancers among males [SIR = 0.32 (95% CI 0.12–0.70)] was observed. The authors suggested that these associations could be a result of higher sun exposure and lower smoking prevalence in the cohort of patients infected by *C. jejuni* compared to general Swedish population, and proposed a speculative hypothesis that smoking may protect from *C. jejuni* infection or affect its symptoms. Anyway, the true association cannot be ruled out despite the weakness of its biological arguments.

Based on the results of only three published studies it is impossible to determine whether *C. jejuni* is a causative agent of cancer. Undoubtedly, the question is open. The design of the study used by Brauner et al. (2010) might be close to ideal, and further investigations, possibly, should use it to reach the objective results.

Fundamental studies on cell and animal models can shed light on the problem as well. At the moment, it is only admissible to include *C. jejuni* in the extended list of the infectious agents that may be a cause of cancer development but are not included in the IARC roster.

3.1.4 Clostridium spp.

The first study that investigated the association of *Clostridium* spp. with cancer was by Mizutani and Mitsuoka (1979) who assessed the effect of intestinal microflora on liver tumorigenesis on gnotobiotic C3H/He male mice monoassociated, diassociated, or polyassociated with the *Escherichia coli*, *Streptococcus faecalis*, *Bifidobacterium adolescentis*, *Bifidobacterium infantis*, *Clostridium* (*C.*) *indolis*, *C. paraputrificum*, *C. perfringens*, *C. innocuum*, *C. nexile*, *C. ramosum*, *C. clostridiiforme*, *Bacteroides multiacidus*, *Bacteroides fragilis*, *Veillonella alcalescens*, *V. parvula*, and *Lactobacillus acidophilus*. The prevalence of liver tumors in this study was higher in most of the gnotobiotes (67–100%) and conventionalized mice (82%) derived from the germfree mice than in the germfree mice (39%). The average number of tumor nodules in gnotobiotes associated with *E. coli*, *S. faecalis*, and *C. paraputrificum* was 2.9. This was significantly higher than in the conventionalized animals (1.6), suggesting that the presence of certain intestinal bacteria is related to hepatocarcinogenesis in gnotobiotic C3H/He male mice. In another article published a year later by Mizutani and Mitsuoka (1980), these authors demonstrated that this promoting effect was suppressed when *Bifidobacterium longum*, *Lactobacillus acidophilus*, and *Eubacterium rectale* were added. Legakis et al. (1981) studied feces from ten colon cancer patients and ten control subjects and revealed significantly higher frequency of isolation and counts of *Bacteroides* species (especially *B. fragilis* group) and *C. paraputrificum* in the patient group. After nearly two decades, Nakamura et al. (2002) studied activities of four microbial enzymes (azoreductase, nitroreductase, beta-glucuronidase, and beta-glucosidase) in major anaerobic members of human fecal microflora and observed that *C. paraputrificum* and *C. clostridiiforme* showed much higher activities than other indicated fecal anaerobes. In particular, nitroreductase activity in *C. paraputrificum* isolated from fecal specimens of patients with colon cancer was significantly higher than in the clostridia isolated from healthy subjects and the subjects given high beef diets. The activities of tested enzymes showed marked differences in each strain of bacteria.

To date, there is definitely insufficient amount of information for the assessment of the *Clostridium* spp. significance in cancer etiopathogenesis. Particular attention should be directed to the *C. paraputrificum* and *C. clostridiiforme* since possible associations of these two bacteria with colon cancer were shown in epidemiological studies as well as in a fundamental study on the animal model. Apparently, it is permissible to include *C. paraputrificum* and *C. clostridiiforme* in the extended list of the infectious agents that may be a cause of cancer development but are not included in the IARC roster.

3.1.5 *Bacteroides spp.*

The investigation of the role of *Bacteroides* in cancer etiology was started by Chung et al. (1975) who examined 23 strains of intestinal anaerobes for indole production from tryptophan. Three isolates of *B. fragilis ss. thetaiotaomicron* and one *Citrobacter* sp. were indole positive. The tryptophanase of the indole-positive strains of intestinal anaerobes was inducible by tryptophan and was susceptible to glucose repression. The tryptophan concentration and tryptophanase activity in feces from rats on an all-meat diet were significantly higher than in feces from rats on a normal diet, suggesting that a population with a diet rich in meat has a greater chance for exposure to possible carcinogens such as indole and other tryptophan metabolites. In the same year, Wheeler et al. (1975) showed that *Clostridium* sp., *C. perfringens*, *Peptostreptococcus productus I*, *Bacteroides fragilis ss. thetaiotaomicron* and *ss. vulgatus* can participate in the metabolic conversion of N-hydroxy-4-acetyl-aminobiphenyl (N-OH-AABP) to carcinogenic 4-acetylaminobiphenyl. Cecal flora was more active in reducing N-OH-AABP than homogenates of tissues of the gastrointestinal tract, and cecal flora also had a greater activity for reducing N-OH-AABP than the stomach flora. Glucuronide conjugates of N-OH-AABP were found in the cecal contents and feces of the germ-free rats only, whilst 4-acetylaminobiphenyl was detected in the feces of conventional rats only. Therefore, the authors suggested that colonic microbiota may affect the level of metabolites of 4-acetylaminobiphenyl which are critical for carcinogenesis. Van Tassel et al. (1982) screened 40 species of anaerobes for their ability to produce an ether-extractable mutagen which was detected in the feces of 15–20% of individuals in populations at high risk for colon cancer. *B. thetaiotaomicron*, *B. fragilis*, *B. ovatus*, *B. uniformis*, and *Bacteroides group 3452A* produced 5- to 8-fold increases in the concentration of mutagen. Mutagen production was constitutive, occurred during the stationary phase of growth, and cell-free extracts produced mutagen faster than whole cells. They proposed that excretion of this mutagen is dependent on the presence of some precursor of unknown origin. López-Boado et al. (2000) detected that acute infection of human colon, bladder, and lung carcinoma cells, primary human tracheal epithelial cells, and human tracheal explants with type 1-piliated *E. coli* mediated a 25–50-fold and long-termed (>24 h) induction of matrilysin (MMP-7) production. Bacterial infection led to the activation of the zymogen form of the enzyme, which was selectively released at the apical surface. Induction of matrilysin was mediated by a soluble, non-LPS bacterial factor and correlated with the release of defensin-like bacteriocidal activity. The bacteria did not induce matrilysin in other cell types, and the expression of other metalloproteinases by epithelial cells was not affected by bacteria. Matrilysin was not detected in germ-free mice, but the enzyme was induced after colonization with *B. thetaiotaomicron*. According to the data of Kim et al. (2002), *B. fragilis* enterotoxin (BFT) elevated NF-κB signaling in HT-29 and T84 epithelial cell lines as well as primary human colon epithelial cells but decreased the signals of IκBα and IκBε. On the contrary, NF-κB inhibition suppressed the BFT-induced neutrophil transepithelial migration in T84 cells, suggesting that

NF-κB can be a central regulator of chemokine gene expression in BFT-stimulated intestinal epithelial cells and may be an important regulator of neutrophil migration. Other metalloprotease produced by *B. fragilis*, fragilysin, hydrolyzes the extracellular domain of E-cadherin, disrupting intercellular adhesion and increasing permeability of the epithelium that, in turn, causes intracellular redistribution of actin with morphologic changes to the cells and release of beta-catenin, which translocates to the nucleus and ultimately increases cellular proliferation. Goodwin et al. (2011) demonstrated that purified BFT upregulates spermine oxidase (SMO), the apolyamine catabolic enzyme highly inducible by inflammatory stimuli, resulting in increased ROS generation and DNA damage in HT29/c1 and T84 colonic epithelial cells. Furthermore, ETBF-induced colitis in C57BL/6 mice was associated with increased SMO expression and treatment of mice with an inhibitor of polyamine catabolism, N(1), N(4)-bis(2,3-butandienyl)-1, 4-butanediamine (MDL 72527), significantly reduced ETBF-induced chronic inflammation and proliferation. In the multiple intestinal neoplasia (Min) mouse model, treatment with MDL 72527 decreased ETBF-induced colon tumorigenesis by 69%.

As mentioned above, Legakis et al. (1981) studied feces from ten colon cancer patients and ten control subjects and revealed significantly higher frequency of isolation and counts of *Bacteroides* species (especially *B. fragilis* group) and *C. paraputrificum* in the patient group. Mikamo et al. (1993) assessed the bacterial flora in 25 patients with cervical cancer and observed that with the progression of cervical cancer, *G. vaginalis*, *B. distasonis* and *P. bivia* were detected at higher rates and higher counts than other bacteria. Moore W.E. and Moore L.H. (1995) found increased counts of *Bacteroides vulgatus*, *Eubacterium* spp., *Ruminococcus* spp., *Streptococcus hansenii*, *Bifidobacterium* spp. and *Faecalibacterium prausnitzii*, and decreased concentration of *Lactobacillus S06* and *Eubacterium aerofaciens* in the feces of patients with polyposis. The first study that demonstrated an increased prevalence of ETBF in colorectal cancer patients was by Toprak et al. (2006) who investigated the prevalence of enterotoxigenic *B. fragilis* (ETBF) in stool specimens from 73 patients with colorectal cancer and from 59 control patients. Enterotoxin gene (bft) was detected by PCR in 38% of the isolates from colorectal cancer patients compared with 12% of the isolates from the control group. Wu et al. (2009) revealed that both enterotoxigenic *B. fragilis* (ETBF) and nontoxigenic B. fragilis (NTBF) chronically colonized mice, but only ETBF triggered colitis and induced colonic tumors in multiple intestinal neoplasia (Min) mice. ETBF induced robust, selective colonic signal transducer and activator of transcription-3 (Stat3) activation with colitis characterized by a selective T helper type 17 (Th17). Antibody-mediated blockade of IL-17 as well as the receptor for IL-23, inhibited ETBF-induced colitis, colonic hyperplasia and tumor formation. Sobhani et al. (2011) collected stool bacterial DNA prior to colonoscopy from 60 colorectal cancer cases and 119 subjects with normal colonoscopy, and revealed significantly higher concentration of *Bacteroides* spp. in colorectal cancer group using pyrosequencing and qPCR. Shen et al. (2010) evaluated adherent bacteria in normal colonic mucosa of 21 adenoma and 23 non-adenoma subjects, revealing lower concentration of *Bacteroidetes* and *Bacteroides* spp. in case group. These results were further proven in the study of Wang et al. (2012)

who recruited 46 patients with colorectal cancer and 56 healthy controls. They detected higher concentration of *Bacteroides* and increased prevalence of *B. fragilis* in the cases that they analyzed whiles *B. uniformis* and *B. vulgatus* were more frequent in controls. Chmelar and Vrtný (2010) determined endotoxin production in 140 strains of the *B. fragilis* group (*B. fragilis*, *B. thetaiotaomicron*, *B. distasonis*, *B. vulgatus*) isolated from patients with 7 cancerous and non-cancerous diseases and detected that all of these bacterial species produced significantly more endotoxins in cancer than in non-cancerous diseases. *B. fragilis* and *B. vulgatus* produced significantly more of endotoxin than *B. distasonis* and *B. thetaiotaomicron.*

To sum up, there are a number of investigations on cell lines, animal models, and population studies which approve the hypothesis of the association of *Bacteroides* spp. (in particular enterotoxigenic *B. fragilis*) with higher colorectal cancer risk. Therefore, it is reasonable to include *B. fragilis* in the short list, and, possibly, other members of *Bacteroides* spp. in the extended list of the infectious agents that may be a cause of cancer development but are not included in the IARC roster.

3.1.6 Salmonella typhi

Typhoid fever, which is a systemic infection caused by Salmonella enterica serovar Typhi (*S. typhi*), is a major health problem in developing countries. There are approximately 21.6 million cases of typhoid fever worldwide and an estimated 200,000 deaths every year. It is known that *S. typhi* may colonize the gallbladder, causing the setting of chronic inflammation. Welton et al. (1979) were the very first to an association between the typhoid-carrier state and death due to malignancies of the hepatobiliary tract. They recruited 471 U.S. carriers of *Salmonella* (*S.*) *typhi*, matched them with 942 controls and demonstrated that chronic typhoid carriers died of hepatobiliary cancer six times more often than the controls. Eleven years later, a case of *S. enteritis* infection which led to a chronic typhoid carrier state, recurrent cholangitis, then sclerosing cholangitis, and, finally, subsequent development of cholangiocarcinoma was reported by Robbins et al. (1988). El-Zayadi et al. (1991) conducted an epidemiological study in Egypt to investigate the role of *Salmonella* in the etiology of bile duct cancer. This research group found 29 consecutive patients who had a bile duct carcinoma, and revealed that 9 out of 23 patients (39.1%) with bile duct carcinoma proved to be *S. typhi/paratyphi* A and B carriers, suggesting that fecal typhoid carrier state was significantly associated with bile duct carcinoma. Caygill et al. (1995) assessed the risk of cancer at various sites in a cohort of chronic carriers of *S. typhi/paratyphi* and observed a greatly increased risk of cancers of the biliary tract, colorectal, pancreatic, lung cancer, and cancer overall. Strom et al. (1995) analyzed the association of typhoid on cancer risk, collecting the medical information from 84 Mexican and Bolivian patients with gallbladder cancer, 264 subjects with cholelithiasis or choledocholithiasis, and from 126 control subjects without stones and with/without cancer. The odds ratio of gallbladder cancer development in individuals with physician-diagnosed typhoid was 12.7 (95% CI=1.5–598).

Nath et al. (1997) collected 1,001 bile specimens from 28 Indian patients with carcinoma of the gallbladder, 56 cholelithiasis cases and 17 individuals without biliary pathology, and detected the higher frequency of *S. typhi/paratyphi-A* in patients with carcinoma of the gallbladder compared to cholelithiasis subjects and controls. Dutta et al. (2000) attempted to investigate the association of chronic *S. typhi* infection with risk of the gallbladder cancer, comparing the frequency of chronic typhoid carriage in the case group (37 Indian patients suffered from carcinoma of the gallbladder along with gallstones) and in the control group (80 subjects with gallstones), 16% of cases were detected to be typhoid carriers, in contrast to 2.5% among controls. The authors identified the typhoid carrier state as an independent risk factor of gallbladder cancer development among patients with cholelithiasis (OR = 14; 95% CI 2–92). In the same year, Shukla et al. (2000) carried out a case-control study to detect typhoid carrier state among the Indian patients with biliary diseases and healthy controls. Subjects with gallbladder carcinoma (29.4%) were more often characterized by *S. typhi*-positivity compared to patients with cholelithiasis (10.7%, OR = 3.86) and controls (5%, OR = 7.19). Totally, there was 8.47 times more risk of gallbladder cancer in culture-positive typhoid carriers than in noncarriers. In the epidemiological investigation carried out in Chili, Serra et al. (2002) recruited 114 Chilean gallbladder cancer cases and 114 matched hospital patients with gallstones, using conditional logistic regression analysis but did not find any association of typhoid history with increased risk of gallbladder cancer in the future. Pandey and Shukla (2003) performed a case-control study in India involving 64 newly diagnosed cases of gallbladder cancer and 101 cases of cholelithiasis. A past history of typhoid was noted in 22% of cancer patients and 13% of gallstone subjects, but the association was not statistically significant. Yagyu et al. (2004) followed-up 113,394 subjects in Japan for 11 years and registered 116 deaths (46 males, 70 females) from gallbladder cancer. After the analysis of risk factors of death from gallbladder cancer, no significant correlation with typhoid was found. However in their meta-analysis and review, Randi et al. (2006) postulated that all the studies based on biological markers, such as serum Vi antigen or the identification of the bacteria in the bile specimens significantly positive association between *S. typhi/paratyphi* carriage and gallbladder cancer risk was observed, and no statistically significant correlation was found only in the studies that used the method of self-reporting. The RRs in the case-control studies ranged between 7.9 (95% CI: 1.7–37.1) and 14 (95% CI: 1.8–92). The cohort study of Caygill et al. showed a RR of 167 (95% CI: 54.1–389) among chronic carriers of the bacteria, but this finding was based on only five gallbladder cancer cases. The summary RR for typhoid infection was 4.8 (95% CI: 1.4–17.3) that became 10.2 (95% CI: 2.0–50.9) after the studies based on self-reported diagnosis of infection was excluded, and the summary RR for case-control studies was 2.6 (95% CI: 1.1–6.1), but became 5.2 (95% CI: 2.1–12.7) after the studies based on self-reported diagnosis of infection was excluded.

Nath et al. (2008) collected hepatobiliary specimens from 52 Indian patients with gallbladder cancer, 223 patients with benign gallbladder diseases, 508 healthy individuals, and 424 corpses without gallbladder diseases. In this investigation,

35 out of the 52 (67.3%) gallbladder cancer cases were PCR-positive for the *S. typhi* flagellin gene that was significantly higher than for patients with benign gallbladder diseases (95/223, 42.6%). In addition, 20/52 (38.5%) of gallbladder cancer patients had significant anti-Vi antibody titers in their serum in comparison with 31/223 (13.9%) for patients with benign gallbladder diseases, and 47/508 (9.2%) for healthy individuals. Thereby, specific nested-PCR amplification of the *S. typhi* flagellin gene in hepatobiliary specimens was more sensitive for the detection of *S. typhi* carriage than anti-Vi antibody titers in serum. In the study of Tewari et al. (2010) 54 Indian patients with gallbladder cancer were matched with 54 patients with gallstone disease (controls). Of the 54 cases, 24 (44.44%) were positive on the Widal test and 12 (22.22%) on indirect hemagglutination assay, compared to 13 (24.07%) and 5 (9.26%) respectively in the controls. 18 (33.33%) cases showed a positive result on PCR (tissue) and 2 on PCR (bile), but none of the controls was positive. Therefore, *S. typhi* was significantly associated with gallbladder cancer, and PCR appeared to be the most specific diagnostic tool. Finally, Safaeian et al. (2011) compared serum *Salmonella enterica serovar typhi* antibody titers between biliary tract cancer cases, biliary stone cases without evidence of cancer, and healthy subjects in China. Their study sample included 627 newly diagnosed primary biliary tract cancer patients; 1,037 biliary stone cases (774 gallbladder and 263 bile-duct) and 959 healthy subjects. The prevalence of *S. typhi* was 1/457 (0.22%), 4/977 (0.41%), and 1/859 (0.12%) among cancer cases, biliary-stone cases, and population controls, respectively, so the authors did not find any statistically significant association (Table 3.3).

To summarize, the association of chronic *S. typhi* carriage with bile duct/gallbladder cancer seems to be plausible, although in some studies no correlation was observed. Firstly, the positive results obtained in almost all studies based on biological markers and substantial value of relative risk calculated in the meta-analysis suggests the true connection. The discrepancies in results between the only biological marker-based study contradictory to others (Safaeian et al., 2011) and investigations performed earlier may be explained by the insufficient sensitivity of the method (serology versus PCR) and/or by extremely low prevalence of *S. typhi*-infection in Chinese population (of note, there were no previous studies in this population devoted to the same problem). Secondly, the rather high number of investigations carried out, the diversity of examined populations (Indian, Mexican, Bolivian, U.S., Egyptian, British), and the big size of the recruited samples testify the hypothesis of the relation of chronic *S. typhi* carriage to gallbladder cancer (and, possibly, to bile duct cancer as well). The absence of basic research analyzing the carcinogenic properties of *S. typhi* does not allow placing it in the short list of the infectious agents that may be a cause of cancer development but are not included in the IARC roster, but this bacterium undoubtedly should be in the extended list. If the fundamental studies on cell lines and animal models will prove the results of epidemiological investigations, *S. typhi* can be placed in the short list.

Table 3.3 The connection between *S. typhi* and bile duct/gallbladder cancer in epidemiological studies

Authors, year, population	Sample size	Methods of detection	Share of positive individuals, OR, 95% CI
Strom et al. (1995), Mexican and Bolivian population	84 cases, 264 controls with cholelithiasis or choledocholithiasis, 126 healthy controls	Self-reporting, serological tests	12.7 (1.5–598)
Singh et al. (1996), Indian population	38 cases, 67 cholelithiasis controls	Detection of *S. typhi* and S. *paratyphi* in bile	13% (cases), 3% (controls) 4.9 (0.9–26.8)
Nath et al. (1997), Indian population	28 cases, 56 cholelithiasis controls, 17 controls with non-biliary diseases	Detection of *S. typhi* and S. *paratyphi-A* in bile	14% (cases), 2% (controls) 9.2 (1.0–86.4)
Dutta et al. (2000), Indian population	37 cases, 80 controls with gallstones	Vi-antigen detection	16% (cases), 2.5% (controls) 14 (2–92)
Shukla et al. (2000), Indian population	51 cases, 40 healthy controls	Vi-antigen detection	29.4% (cases). 10.7% (controls with cholelithiasis), 5% (healthy controls) 3.86 compared to cholelithiasis 7.19 compared to controls
Serra et al. (2002), Chilean population	114 cases, 114 controls with gallstones	Self-reporting	10% (cases), 8% (controls) 0.5 (0.2–1.2)
Pandey and Shukla (2003), Indian population	64 cases, 101 controls with cholelithiasis	Self-reporting	22% (cases), 13% (controls) 1.3 (0.9–2.0), non-significant
Yagyu et al. (2004), Japanese population	33 cases, 38,608 controls	Self-reporting	6% (cases), 1.5% (controls) 2.1 (0.5–8.7), non-significant
Nath et al. (2008), Indian population	52 cases, 223 controls with benign gallbladder diseases, 508 healthy controls, 424 corpses without gallbladder diseases	PCR	67.3% (cases), 42.6% (controls with benign gallbladder diseases), 8.2% (corpses) OR = 2.37 (compared to controls with benign gallbladder diseases) OR = 22.9 (compared to corpses without gallbladder diseases)
Tewari et al. (2010), Indian population	54 cases, 54 controls with gallstones	Widal test. PCR	44.4% (cases), 24% (controls) 33.33% (cases), 0% (controls)
Safaeian et al. (2011), Chinese population	627 cases, 1,037 biliary stone controls (774 gallbladder and 263 bile duct) and 959 healthy controls	Vi-antigen detection	0.22% (cases), 0.41% (biliary-stone controls), 0.12% (healthy controls)

3.1.7 *Streptococcus spp.*

Streptococcus bovis (*S. bovis*) is a catalase- and oxidase-negative, non-motile, non-sporulating, Gram-positive lactic acid bacterium that grows as pairs or chains of cocci. *S. bovis* is commonly found in the alimentary tract of cows, sheep, and other ruminants. In addition, it is a known human pathogen colonizing the gut and being implicated as a causative agent of endocarditis, neonatal septicemia and meningitis. The association between *S. bovis* and colorectal cancer were observed by Klein et al. (1977) who isolated *S. bovis* from fecal cultures of 35 of 63 patients with carcinoma of the colon, 4 of 21 with non-colonic neoplasms, 7 of 25 with inflammatory bowel disease, 5 of 37 with other gastrointestinal disorders, and 11 of 105 controls. The prevalence of *S. bovis* in fecal cultures from colorectal cancer patients was significantly increased compared to the controls and all other groups, but these results were not confirmed in the subsequent studies of Norfleet et al. (1993) and Potter et al. (1998). These authors neither cultured *S. bovis* more frequently from cancerous tissues than from non-cancerous nor demonstrated increased fecal carriage rate of this bacterium in colorectal cancer cases. In the following years, Burns et al. (1985) found that patients with colonic carcinoma and subjects with polyps that are at increased risk for malignant degeneration were characterized by a statistically significant increase in the fecal carrier rate for *S. bovis* over the group with benign colon, and Arzanauskiene et al. (2003) revealed that the prevalence of *S. bovis* in fecal cultures from patients with carcinoma of the colon is significantly increased (56%) in comparison with cultures from patients with inflammatory bowel diseases (28%) and from healthy controls (10%). In addition, the association of bacteremia and infectious endocarditis caused by *S. bovis* infection with colorectal cancer was further demonstrated in a lot of case series reported by Murray and Roberts (1978), Levy and Brooks (1978), Klein et al. (1979), Wilson et al. (1981), Reynolds et al. (1983), Beeching et al. (1985), Leport et al. (1987), Pigrau et al. (1988), Ruoff et al. (1989), Zarkin et al. (1990), Hoen et al. (1994), Ballet et al. (1995), Kupferwasser et al. (1998), Gonzalez-Quintela et al. (2001), Duval et al. (2001), Pergola et al. (2001), Herrero et al. (2002), González-Juanatey et al. (2003), Lee et al. (2003), Tripodi et al. (2004), Gold et al. (2004), Jean et al. (2004), Corredoira et al. (2005), Alazmi et al. (2006), Giannitsioti et al. (2007), Corredoira et al. (2008a, 2008b), Vaska and Faoagali (2009), Fernández-Ruiz et al. (2010) and case reports published by Steinberg and Naggar (1977), Noble et al. (1978), Smith et al. (1978), Dunham et al. (1980), Weitberg et al. (1981), Friedrich et al. (1982), Hossenbux et al. (1983), Robbins and Klein (1983), Kewal et al. (1983), Belinkie et al. (1983), Silver (1984), Trajber et al. (1984), Dowd and Loftus (1985), Friedman et al. (1986), Tabibian and Clarridge (1989), Devis et al. (1989), McMahon et al. (1991), Almeida et al. (1992), Ma et al. (1992), Copeland and Malster (1993), Ackerman et al. (1995), Cailliez-Tomasi et al. (1997), Goumas et al. (1997), Mann et al. (1997), Genta et al. (1998), Seglenieks and Black (1998), Garcia-Porrua et al. (2000), Waisberg et al. (2002), Vince et al. (2003), Montalto et al. (2004), Weng et al. (2005), Beadsworth et al. (2005), Aizawa et al. (2006), Wentling et al. (2006), Kok et al. (2007), Apsingi et al. (2007), Harnik (2007), Bleibel et al. (2007),

Ferrari et al. (2008), Hernandez-Vaquero et al. (2008), Seifert (2009), Florescu et al. (2009), Gupta et al. (2010), Kim et al. (2010a, b), Barcelos et al. (2010), Srivastava et al. (2010), Liddle et al. (2012), Maurya (2011). Besides, some other *Streptococcus* species (*S. milleri*, *S. salivarius*, *S. intermedius*, *S. anginosus*, *S. sanguis*) were found to be associated with some malignancies in a number of case reports published by Gleeson et al. (1983), Legier (1991), Smit et al. (1994), Gil Lasa et al. (1995), Kampe et al. (1995), Siegert and Overbosch (1995), Buckley et al. (1998), Macaluso et al. (1998), Conde Martel et al. (2002), Tzur et al. (2003), Afek et al. (2004), Millichap et al. (2005), Stojan et al. (2006), Ballardini et al. (2008), Lin et al. (2008), Kwon et al. (2009), Nijjer and Dubrey (2010). Tjalsma et al. (2006) developed the immunocapture mass spectrometry assay able to generate *S. bovis* infection-related protein profiles. The composition of these profiles was governed by the capture of specific antigens by serum antibodies from colon cancer patients. The assay revealed that *S. bovis* antigen profiles could distinguish 11 out of 12 colon cancer patients from 8 control subjects. Moreover, *S. bovis* antigen profiles were also detected in polyp patients, indicating that infection with this bacterium does occur early during carcinogenesis. One of the diagnostic antigens was a surface-exposed heparin-binding protein, which can be involved in *S. bovis* attachment to tumor cells. Abdulamir et al. (2009) measured IgG antibodies of *S. gallolyticus* in sera of 50 colorectal cancer cases, 14 colorectal adenoma patients, 30 age- and sex-matched apparently healthy volunteers and 30 age- and sex-matched colonoscopically-proven tumor-free control subjects. In their study, colorectal cancer and adenoma patients were associated with higher levels of serum *S. gallolyticus* IgG antibodies in comparison with healthy volunteers and control subjects. In addition, the expression of NF-κB and IL-8 mRNA was higher in *S. gallolyticus* IgG-seropositive than in seronegative groups in both tumorous and non-tumorous sections. The authors concluded that *S. gallolyticus* plays an essential role in the oncogenic progression of normal colorectal mucosa to adenoma and then to colorectal cancer. Boleij et al. (2010) assessed the presence of serum antibodies against S. bovis antigen RpL7/L12 in Dutch population (82 cases, 127 controls) and American population (64 cases, 48 controls) using ELISA, and found that this bacterial antigen was increased in polyp patients and stage I/II colorectal cancer patients compared to asymptomatic individuals, suggesting that *S. bovis*-infection may be a risk factor for the progression of premalignant lesions into early stage carcinomas. Two years later, the same research group identified increased serum antibody titers to *S. bovis* biotype I pilus proteins as a potential biomarker for diagnosis of colorectal cancer cases associated with this bacterium (Boleij et al., 2012). Meta-analysis of Boleij et al. (2011a) revealed that among the *S. bovis*-infected patients who underwent colonic evaluation, the median percentage of patients who had concomitant adenomas/carcinomas was 60%, which largely exceeded the disease rate reported in the general asymptomatic population. Patients with *S. bovis* biotype I (*S. gallolyticus* subsp. *gallolyticus*) infection were characterized by a strongly increased risk of having colorectal cancer (pooled OR = 7.26; 95% CI = 3.94–13.36) compared with *S. bovis* biotype II-infected patients. Colorectal cancer occurred more often among patients with *S. bovis* infectious endocarditis than among patients with *S. bovis* infection at other sites (pooled OR = 3.72, 95% CI = 2.03–6.81).

Sasaki et al. (1995, 1998) investigated tissue samples from 15 patients with esophageal cancer, 43 gastric cancer subjects, 16 lung cancer subjects, 10 cervical cancer cases, 14 patients with renal cell carcinoma, 10 colorectal cancer cases, and 19 bladder cancer subjects. They frequently found *S. anginosus* DNA sequences in samples from esophageal cancer and gastric cancer tissues whereas *S. anginosus* DNA bands were not detected in the noncancerous portions of the esophagus or the stomach. In addition, PCR analysis revealed *S. anginosus* sequences only in 7% of the noncancerous portion of the esophagus, suggesting that this bacterium may play a role in the etiopathogenesis of these cancer types. However, these data were not confirmed by Morita et al. (2005) who did not reveal statistically significant differences in level of *S. anginosus* comparing 41 esophageal cancer patients and 22 healthy people. Furthermore, Tateda et al. (2000) analyzed 217 DNA samples prepared from head and neck squamous cell carcinomas and detected *S. anginosus* DNA sequence by PCR in all of them. These results were then confirmed in the investigation of Shiga et al. (2001). Morita et al. (2003) found that 8 of 18 samples (44%) from the esophageal cancer patients and 5 of 38 samples (13%) of oral cancer cases contained a detectable level of *S. anginosus* DNA, and the quantity of *S. anginosus* DNA in the esophageal cancer tissues was significantly higher than in oral cancer, so *S. anginosus* was associated with esophageal cancer, but not with oral cancer. On the contrary, most non-cancerous tissues of the esophagus and tongue showed an undetectable level of *S. anginosus*. Narikiyo et al. (2004) examined the bacterial diversity in esophageal carcinoma tissues and in the saliva of healthy people. This group reported the preferential and frequent infection of the *S. mitis* and *S. anginosus* in esophageal cancer from different regions of the world (for *S. anginosus*, the frequency was 69% in Japanese sample and 91% in sample consisted of Chinese, French, and Italian patients). *S. anginosus* and *S. mitis* induced the production of procarcinogenic IL-8 and GRO-α by esophageal tissues. So, these two bacteria could have significant role in the carcinogenic processes in the esophagus. Sasaki et al. (2005) collected tissue specimens from 46 oral cancer and three precancerous leukoplakia subjects and found *S. anginosus* DNA in 19 of 42 samples of squamous cell carcinoma, but not in other types of cancer (lymphoma and rhabdomyosarcoma) or leukoplakia samples, suggesting that this bacterium may be involved in oral squamous cell carcinoma development. Mager et al. (2005) analyzed 45 cases of oral squamous cell carcinoma and 229 healthy controls, revealing that *Capnocytophaga gingivalis*, *Prevotella melaninogenica* and *Streptococcus mitis* were elevated in the saliva of individuals with oral squamous cell carcinoma compared to controls.

Ellmerich et al. (2000) revealed that the administration of either *S. bovis*orits cell wall antigens promoted the progression of intestinal preneoplastic lesions through the increased formation of hyperproliferative aberrant colonic crypts, enhanced the expression of proliferation markers and increased the production of IL-8 in the colonic mucosa. In addition, in this investigation bacterial wall proteins were more potent inducers of neoplastic transformation than the intact bacteria. Biarc et al. (2004) showed a relationship between the pro-inflammatory potential of *S. bovis* proteins and their procarcinogenic properties (induction of IL-8, PGE2 and COX-2 expression, promotion of neoplastic lesions) confirming the linkage between inflammation and colon carcinogenesis. Sugano et al. (2003) detected the significant elevation in carcinogenic 8-hydroxy-deoxyguanosine (8-OHdG, resulting from the continuous generation of reactive oxygen species associated with chronic inflammation) in patients who tested positive for *S. anginosus* than in patients who tested negative for the bacterium. Kurkivuori et al. (2007) found that the clinical strain of *Streptococcus salivarius*, both clinical and culture collection strains of *Streptococcus intermedius* and culture collection strain of *Streptococcus mitis* produced high amounts of acetaldehyde and showed significant alcohol dehydrogenase activity, suggesting that viridans group streptococci may participate in metabolizing ethanol to carcinogenic acetaldehyde in the mouth. Boleij et al. (2011b) reconstructed the route of *S. bovis* biotype I in vitro with adhesion, invasion, and translocation assays on differentiated Caco-2 cells. In this research, *S. gallolyticus* subsp. *gallolyticus* strains had a relative low adhesiveness and could not internalize epithelial cells but were able to paracellularly cross a differentiated epithelium without inducing epithelial IL-8 or IL-1β responses and had an outstanding ability to form biofilms on collagen-rich surfaces, which in vivo are found in damaged heart valves and (pre) cancerous sites with a displaced epithelium.

To conclude, there are basic as well as extensive epidemiological evidence that *S. bovis biotype I* (*S. gallolyticus* subsp. *gallolyticus*) may be a cause of colorectal cancer. According to our criteria, it is possible to place it in the short list of the infectious agents that may be a cause of cancer development but are not placed in the IARC roster. It is also possible to place *S. anginosus* in the extended list of the infectious agents that may be a cause of cancer development but are not placed in the IARC roster.

3.2 The Role of Bacteria in Cancer of the Respiratory Tract

3.2.1 Mycobacterium tuberculosis

Tuberculosis (TB), a destructive disease afflicting the lungs, caused by *Mycobacterium tuberculosis* (*M. tuberculosis*) is a major global health burden, with about nine million of new cases and 1.1 million deaths annually. When the host protective immunity fails to control *M. tuberculosis* growth, progression to active disease occurs. When the immune system successfully prevents the infection (90% of infected individuals), the disease becomes latent, and reactivation of the active disease may occur later in life when there is a depression of the host immunity.

According to the data of the last comprehensive systematic review and meta-analysis published by Brenner et al. (2011), there were 30 studies (Hinds et al. 1982; Samet et al. 1986; Gao et al. 1987; Zheng et al. 1987; Wu et al. 1988; Wu-Williams et al. 1990; Alavanja et al. 1992; Ger et al. 1993; Wu et al. 1995; Schwartz et al. 1996; Wang et al. 1996a, b; Ko et al. 1997; Brownson and Alavanja 2000; Osann et al. 2000;

Cocco et al. 2000; Brenner et al. 2001; Le et al. 2001; Kreuzer et al. 2001, 2002; Chan-Yeung et al. 2003; Zatloukal et al. 2003; Littman et al. 2004a, b; Ramanakumar et al. 2006; Galeone et al. 2008; Engels et al. 2009; Wang et al. 2009; Liang et al. 2009; Brenner et al. 2010; Koshiol et al. 2010) conducted in North America, Europe and Asia, which investigated the association of tuberculosis on lung cancer risk with adjustment for smoking. The relative risk (RR) of lung cancer development among patients with TB history was 1.76 (95% CI = 1.49–2.08). This elevation was similar for all studies when compared to only never smokers, and TB risk among never-smokers was slightly increased (RR = 1.90, 95% CI = 1.45–2.50). Besides, a history of pulmonary tuberculosis is an independent risk factor for lung cancer survival, according to the data of Heuvers et al. (2012). In addition, Nalbandian et al. (2009) presented experimental evidence of lung cell dysplasia induction and lung squamous cell carcinoma formation by chronic TB infection in a lung-specific manner. In their research, squamous cell aggregated consistently among the lung tissue associated with chronic TB lesions, in some cases resembling squamous cell carcinomas, and a transplantable tumor was established after the transfer of cells that were isolated from TB lung lesions into syngeneic recipients. *M. tuberculosis*-infected macrophages demonstrated a key role in TB-induced carcinogenesis, inducing DNA damage in their vicinity and producing a potent epidermal growth factor epiregulin, which may serve as a paracrine survival and growth factor responsible for squamous metaplasia and tumorigenesis. Lung carcinogenesis during the course of chronic TB infection was more eminent in animals with severe lung tissue damage mediated by TB-susceptibility locus sst1.

Besides, 22 studies from North America, Europe and East Asia investigated the association between pneumonia and lung cancer risk while adjusting for smoking (Gao et al. 1987; Wu et al. 1988, 1995; Wu-Williams et al. 1990; Alavanja et al. 1992; Schwartz et al. 1996; Laurila et al. 1997;; Jackson et al. 2000; Brownson and Alavanja 2000; Brenner et al. 2001; Kreuzer et al. 2001, 2002; Anttila et al. 2003; Zatloukal et al. 2003; Littman et al. 2004a, b; Schabath et al. 2005; Ramanakumar et al. 2006; Gorlova et al. 2006; Cassidy et al. 2008; Wang et al. 2009; Koshiol et al. 2010). A significant increase in lung cancer risk was observed among all studies (RR = 1.43, 95% CI = 1.22–1.68). The effect was similar for all the combined studies compared to studies with never-smokers only (RR = 1.36, 95% CI = 1.10–1.69). Nevertheless, there is no experimental evidence that all specific pneumonia etiological agents (possibly excluding *Chlamydia pneumoniae*) may cause cancer.

In summary, there is a lot of consistent epidemiological evidence that *M. tuberculosis* may cause lung cancer as defined fundamental carcinogenic properties of this bacterium. According to our criteria, it is possible to place it in the short list of the infectious agents that may be a cause of cancer development but are not placed in the IARC roster. As carcinogenic features of other pneumonia etiological agents excluding *Chlamydia pneumoniae* (*Streptococcus pneumoniae*, *Staphylococcus aureus*, *Klebsiella pneumoniae*, *Mycoplasma pneumoniae*, *Pseudomonas aeruginosa*, *Haemophilus influenzae*, *Legionella pneumophila*, *Pneumocystis carinii*, many types of viral pneumonia etc.) are not determined, it is not possible to place them in the list but, possibly, they can be attractive targets for the fundamental investigations from the point of view of cancer biologist.

3.2.2 *Chlamydia pneumoniae*

Chlamydia pneumoniae (*C. pneumoniae*) is a common intracellular bacterium that causes pneumonia and other respiratory infections world-wide. Like all *Chlamydia*, it has a tendency to cause persistent and chronic infections. *C. pneumoniae* has been serologically associated with chronic lung diseases such as chronic bronchitis, asthma, and atherosclerosis. On the basis of epidemiological evidence suggesting chronic obstructive pulmonary diseases as an independent risk factor for lung cancer, Laurila et al. (1997) assessed the association between chronic *C. pneumoniae* infection and risk of lung cancer among male smokers. They recruited 230 Finnish patients with lung cancer and 230 healthy controls and conducted a seroepidemiological study, revealing markers indicating chronic *C. pneumoniae* infection in 52% of cases and 45% of controls (OR = 1.6, 95% CI = 1.0–2.3). The risk was greater in men younger than 60 years (OR = 2.9, 95% CI = 1.5–5.4) but not in the older age group (OR = 0.9, 95% CI = 0.5–1.6). Their results were further confirmed by Koyi et al. (1999) in a case-control seroepidemiological study of 117 patients with lung cancer who detected specific *C. pneumoniae* IgA antibodies significantly more often in lung cancer cases than in control groups with coronary heart disease and in healthy controls after adjustment for smoking. Jackson et al. (2000) conducted a case-control study of 143 U.S. cases and 147 controls and confirmed the relation of lung cancer risk with anti-*C. pneumoniae* IgG, IgM, and IgA antibodies. In multivariate analysis controlling for smoking variables and educational status, IgA antibody titer 216 was independently associated with risk of lung cancer among subjects <60 years of age (OR = 2.67, 95% CI = 1.21–5.89) but not among older subjects (OR = 0.69, 95% CI = 0.34–1.43). Koyi et al. (2001) recruited 210 consecutive patients with lung cancer during a 2-year period and found that lung cancer cases had higher antibody titers than controls, and male lung cancer patients had significantly higher levels of IgG and/or IgA antibodies than females. Anttila et al. (2003) showed the correlation between antibodies against *C. pneumoniae* and risk of female early-onset lung cancer among Finnish women (58 cases, 287 controls, OR = 1.8). Kocazeybek (2003) examined 123 Turkish lung cancer patients and 123 matched healthy controls, revealing that in male patients with lung carcinoma *C. pneumoniae* IgG antibody titers of ≥512 and IgA antibody titers of ≥40 were found more frequently than in the control group. However, this ratio was not significant for the female patients, and chronic *C. pneumoniae* infections were seen significantly more often in male patients with carcinoma who were aged 55 years or younger. In a matched case-control study of Littman et al. (2004a, b) including 508 pairs of matched U.S. cases and controls, individuals with antibody titers ≥16 had 1.2-foldrisk of lung cancer (95% CI = 0.9–1.6) compared to those with lower titers. There was a significant trend (P = 0.007) of OR elevation with increasing IgA titers (till OR = 2.8, 95% CI = 1.1–6.7) associated with titers > or =256. On the contrary, Koh et al. (2005) created a sample of 233 non-smoking Chinese women with lung cancer and 187 controls and measured anti-*C. pneumoniae* antibody titers in their blood, but did not detect the statistically significant association of them with lung cancer risk. Littman et al. (2005) noted that in studies in which blood was collected before diagnosis, were characterized by weaker results (OR range = 1.2–2.1) than studies in which blood was sampled after diagnosis (OR range = 1.4–9.9). ORs for IgA ≥ 16 ranged from 1.2 to 1.7, but when a higher IgA cutoff (from ≥64) was used, the ORs were higher (OR range = 2.1–9.9). In 2008, Smith et al. (2008) conducted a multicentric seroepidemiological study involving six countries (Sweden, Germany, France, Russia, Romania and Italy), by recruiting 163 lung cancer non-smoking cases and 190 controls, but they did not find a statistically significant relationship between anti-*C. pneumoniae* antibodies and lung cancer risk. The authors suggested that smoking could affect the results of studies where the association was observed. Finally, Zhan et al. (2011) recently conducted a meta-analysis in which they found 12 studies, involving 2,595 lung cancer cases and 2,585 controls from 4 prospective studies and 8 retrospective studies. According to their calculations, people exposed to *C. pneumoniae*-infection had an OR = 1.48 (95% CI = 1.32–1.67) for lung cancer risk. *C. pneumoniae*-infection was defined as a risk factor for lung cancer in both prospective studies (OR = 1.16, 95% CI = 1.00–1.36) and retrospective studies (OR = 2.17, 95% CI = 1.79–2.63) and in both IgA ≥ 16 cutoff group (OR = 1.22, 95% CI = 1.06–1.41) and the IgA ≥ 64 cutoff group (OR = 2.35, 95% CI = 1.88–2.93) (Table 3.4).

Concerning other tumors, Anttila et al. (1998) investigated the possible association between chronic chlamydial infections and malignant lymphomas, by carrying out one more seroepidemiological case-control study including 72 patients with lymphoma [53 with non-Hodgkin's lymphoma (NHL) and 19 with Hodgkin's disease (HD)] and 72 matched controls. Again, they found an association between the presence of *C. pneumoniae*-specific immune complexes and NHL (OR = 7.3, 95% CI = 2.2–25) that seemed to be limited to men. When the ratios of *C. pneumoniae* antibodies measured by ELISA were calculated, the mean ratio was significantly higher in NHL cases (M = 4.9, range 0.9–23) and HD cases (M = 4.0, range 0.5–9) than in their controls (M = 1.9, range 0.6–8 and 2.1, range 0.6–9, respectively). Chan et al. (2006) and Shen et al. (2006) reported a case of *C. pneumoniae* DNA identification in the orbital lymphoma of a Chinese patient from Hong Kong. Chanudet et al. (2007) showed that the frequency of *C. pneumoniae* DNA identification in pulmonary MALT lymphoma tissues varies from 2 out of 35 (6%) in U.K. to 3 out of 21 (14%) in Germany and to 3 out of 13 (23%) in Italy. In Italy, it was significantly higher compared to other lymphoproliferative diseases (1 out of 10, 10%). In addition, Moutel et al. (2004) did not detect *C. pneumoniae* DNA in 9 out of 10 head and neck malignant tumors suggesting that this bacterium does not play a major role in the development of this cancer type. Zhang et al. (2009) in China identified *C. pneumoniae* DNA in 2 out of 38 cases (5.3%) of ocular adnexal MALT lymphoma.

Abrams et al. (1999) discovered that Sézary T cell-activating factor (SAF) is a *C. pneumoniae*-associated protein, and 2 years later (Abrams et al. 2001) they showed that the combination of Sézary T cell-activating factor (SAF) and IL-2 stimulated the propagation of T cell lines from the PBMC of patients with Sézary syndrome, demonstrating that one third of those cell lines contain the predominant malignant clone. So, the authors proposed that chronic *C. pneumoniae*-infection results in the expansion of *C. pneumoniae*-specific T cells, potentiating the development of cutaneous

Table 3.4 The connection of *C. pneumoniae* with lung cancer in epidemiological studies

Authors, year, population	Sample size	Methods of detection	Share of positive individuals, OR, 95% CI
Laurila et al. (1997), Finnish population	230 cases, 230 healthy controls	MIF	53% (cases), 45% (controls) 1.6 (1.0–2.3) In men younger than 60 years: 2.9 (1.5–5.4) IgA is the indicator
Koyi et al. (1999), Swedish population	117 cases, controls with coronary heart disease, healthy controls	Serological tests	Positive association detected IgA is the indicator
Jackson et al. (2000), U.S. population	143 cases, 147 controls	MIF	50% (cases), 26% (controls) Among subjects <60 years of age: 2.67 (1.21–5.89) IgA is the indicator
Koyi et al. (2001), Swedish population	210 cases, healthy controls	Serological tests	Positive association detected IgA is the indicator
Kocazeybek (2003), Turkish population	123 cases, 123 controls	MIF	80% (cases), 43% (controls) For male patients: positive association detected IgG and IgA are indicators
Littman et al. (2004a, b), U.S. population	508 cases, 508 controls	MIF	55.4% (cases), 51.2% (controls) 1.2 (0.9–1.6) IgA is the indicator
Koh et al. (2005), Chinese population	233 cases, 187 controls	MIF	No association
Smith et al. (2008), Swedish, German, Russian, French, Romanian, Italian population	163 non-smoking cases, 190 controls	MIF	No association
Zhan et al. (2011), Meta-analysis	2,595 cases, 2,585 controls		1.48 (1.32–1.67) In prospective studies: 1.16 (1.00–1.36) In retrospective studies: 2.17 (1.79–2.63) In IgA ≥ 16 cutoff group: 1.22 (1.06–1.41) In IgA ≥ 64 cutoff group: 2.35 (1.88–2.93)

T cell lymphoma. Carratelli et al. (2002) discovered that *C. pneumoniae* may protect THP-1 cells against the apoptosis which spontaneously occurs in macrophages in the absence of an activation signal, but heat-inactivated bacteria do not give significant protection against apoptosis, while UV-treated Chlamydia did provide a degree of protection against apoptosis. Costa et al. (2002) suggested that chlamydial HSP60 is an important mediator of inflammatory responses during *C. pneumoniae*-infection. Rössler et al. (2003) detected neither *C. pneumoniae*-specific DNA sequences nor antigens in skin biopses from 24 patients with cutaneous T-cell lymphomas. Yang et al. (2003) demonstrated that heparin treatment of *C. pneumoniae* significantly decreased its ability to induce IL-8 and TNF-alpha mRNA in human lung carcinoma cells, indicating that bacterial adhesion is an important early stimulus for induction of proinflammatory mediators. Heat or UV inactivation of *C. pneumoniae* only partially reduced the cytokine response, and inhibition of *C. pneumoniae* protein or DNA synthesis did not affect its ability to induce cytokine gene expression. Zhang et al. (2011) investigated the effect of *C. pneumoniae*-infection on human laryngeal carcinoma cell line HEp-2 cell adhesion and migration, to clarify the role and mechanism of *C. pn* infection in tumor metastasis. At 72 h after infection, *C. pneumoniae* infected-HEp-2 cells were swollen and partially desquamated. Multiple vacuoles were observed and *C. pneumoniae* inclusions occupied the whole cytoplasm of the HEp-2 cells. In the cell adhesion assay, the A value in *C. pneumoniae* infection group was significantly higher than that in the control group at 2 h after infection. The cell adhesion ratio in the *C. pneumoniae* infection group was 119.89%. The migration distance of *C. pn* infected-HEp-2 cells in the wound-healing assay was significantly longer than that of control cells at 24 h after infection, and HEp-2 cells infected with *C. pneumoniae* for 12 h migrated more than the control cells in the transwell assay. The authors concluded that *C. pneumoniae*-infection can promote HEp-2 cell adhesion to collagen I and migration of HEp-2 cells, proposing that *C. pneumoniae*-infection may play a role in the promotion of the metastasis of laryngeal malignant tumors.

To sum up, there are basic as well as extensive epidemiological evidence that *C. pneumoniae* may cause lung cancer, but the fairness of this conclusion may differ in distinct countries, and the results of different studies may be influenced by the smoking status of the study sample. For other malignancies, the number of investigations is too small to establish *C. pneumoniae* as their possible etiological agent. According to our criteria, it is possible to include it in the short list of the infectious agents that may be a cause of cancer development but are not placed in the IARC roster.

3.3 The Role of Bacteria in Cancer of the Genitourinary Tract

3.3.1 Gardnerella vaginalis, Treponema denticola, Bacteroides distasonis, Prevotella bivia, Mycoplasma hominis, Ureaplasma urealyticum

Guijon et al. (1992) examined vaginal microbial flora of 106 women with cervical intraepithelial neoplasia (CIN) and 79 controls and found that human papillomavirus (HPV) infection, *Mycoplasma* (*M.*) *hominis* vaginal infection and abnormal vaginal flora were considerably associated with CIN, suggesting that CIN may be promoted by vaginal micro-organisms in conjunction with HPV-infection. A year later,

Mikamo et al. (1993) assessed the bacterial flora in 25 patients with cervical cancer and observed that with the progression of cervical cancer, *Gardnerella (G.) vaginalis*, *Bacteroides distasonis* and *Prevotella (P.) bivia* were detected at higher rates and higher counts than other bacteria. Neuer and Menton (1995) evaluated the cervical flora in 216 patients with CIN (grades I-III) and in 100 asymptomatic controls. According to their data, *G. vaginalis* combined with *Mycoplasma* spp. and *Bacteroides* spp. were found in 22% of the patients with dysplasia and in 5% of the control group, indicating that women with cervical dysplasia may have a higher prevalence of pathological cervical flora in comparison with healthy controls. Six years after their first study, Mikamo et al. (1999) investigated the intravaginal bacterial flora in 20 patients with cervical cancer and in control group of pre- and post-menopausal women with benign gynecological diseases. They revealed *G. vaginalis* in 50% of cases, bacterial vaginosis was present in 8 of those 10 patients (80%), and the count of *G. vaginalis* detected was higher compared to other coexisting species and was increased in cases compared to controls. Narikiyo et al. (2004) examined the bacterial diversity in esophageal carcinoma tissues and in the saliva of healthy people. This group reported the preferential and frequent infection of the *T. denticola*, *S. mitis*, and *S. anginosus* in esophageal cancer from different regions of the world (for *T. denticola*, the frequency was 38% in Japanese sample and 46% in sample consisted of Chinese, French, and Italian patients). Their data suggested that these three bacteria could have significant roles in the carcinogenic process of many cases of esophageal cancer by causing inflammation and by promoting the carcinogenic process, and that eradication of these three bacteria may decrease the risk of recurrence. Lukic et al. (2006) detected a high association between the prevalence of *Ureaplasma urealyticum* infection and the grade of cytological cervical lesion (27% for ASC-US, 35% for L-SIL and 45% for H-SIL). Only 19% of the controls were positive for this bacterium that was significantly less than that observed in the positive cytology groups. They also observed the association of HPV with *Ureaplasma urealyticum* in patients with H-SIL vaginal cytology (83%) that was much higher than in patients with slightly abnormal or normal vaginal cytology (56% for ASC-US, 49% for L-SIL, 40% for normal cytology). In contrast, the association between HPV and multiple microorganisms decreased with the level of cellular dysplasia in 30% of controls, 33% of ASC-US, 32% of L-SIL and 17% of H-SIL. The authors concluded that the presence of a high *Ureaplasma urealyticum* level seems to be a cofactor of HPV infection. Dicksved et al. (2009) investigated the gastric microbiota from ten patients with gastric cancer and then characterized and compared it with microflora of five dyspeptic controls. The terminal restriction fragment length polymorphism (T-RFLP) analysis revealed a complex bacterial community in the cancer patients that was not significantly different from that in the controls. Sequencing of 140 clones revealed 102 phylotypes, with representatives from five bacterial phyla (*Firmicutes*, *Bacteroidetes*, *Proteobacteria*, *Actinobacteria* and *Fusobacteria*). This research detected a relatively low abundance of *H. pylori* and showed that the gastric cancer microbiota was dominated by different species of the genera *Streptococcus*, *Lactobacillus*, *Veillonella* and *Prevotella*. Ekiel et al. (2009) assessed the occurrence of genital mycoplasmas, particularly *Ureaplasma parvum* and *Ureaplasma urealyticum*, in 143 women with a typical squamous cells of undetermined significance (ASCUS), low grade squamous intraepithelial lesions (LSIL) and high grade squamous intraepithelial lesions (HSIL), in comparison with 39 healthy controls. Urogenital mycoplasmas were detected in women with HSIL significantly more often compared to women with LSIL, ASCUS, and with normal cytology.

Barykova et al. (2010) were the first who tried to investigate the role of *M. hominis* in prostate cancer etiology. They collected 250 samples (from each lobe of prostate obtained from 125 Russian patients with suspected prostate cancer), and found that 20.5% of cases were positive for Mycoplasma by PCR. In addition, out of 118 serum samples, 30.5% were positive for IgG to *M. hominis* in ELISA. In the following year, Barykova et al. (2011) continued their study, detected *M. hominis* by standard PCR in 15% of the 125 patients of the first group and by quantitative real-time PCR in 37.4% of the 123 subjects in the second group (totally, 89 prostate cancer cases, 55 high-grade prostate intraepithelial neoplasia (HGPIN) controls, 104 benign prostatic hyperplasia (BPH) controls, 27 men without detectable prostate disease). The comparison with the control group revealed that *M. hominis* was present three times more frequently in patients with prostate cancer than in those with benign prostatic hyperplasia (BPH). It should be noted that, no *M. hominis* was detected in the prostates of 27 men without detectable prostate disease. In addition, the subjects with prostate cancer had higher titers of antibodies against *M. hominis*, and the average PSA levels were higher in *M. hominis*-positive men. The American sample of patients (27 prostate cancer cases, 31 controls) was too small to obtain statistically significant results, although the positive trend was observed (*M. hominis* was detected in 18.5 and 12.9% of cases and controls, respectively). These data, together with previous observations linking mycoplasma infection with cell transformation, genomic instability and resistance to apoptosis, suggested that *M. hominis* infection may be involved in prostate cancer development. In Czech Republic, Hrbacek et al. (2011) recruited a sample of 329 prostate cancer patients and 105 controls with BPH. In their study, antibodies against *Ureaplasma urealyticum* were detected in prostate cancer cases two times more frequent than in control group (OR=2.06; 95% CI=1.08–4.28) but no significant differences for antibodies against *M. hominis* were found. Conversely, men with BPH were more often seropositive for HPV 18 and *C. trachomatis* (OR=0.23; 95% CI=0.09–0.61 and OR=0.45; 95% CI=0.21–0.99, respectively) and had higher mean serum CMV antibody levels than subjects with prostate cancer. It is contradictory in comparison with the findings of Barykova et al. (2011) who did not reveal *U. urealyticum* in significant share of patients with prostatic diseases (only 0.8%).

Leung et al. (2002) analyzed the effects of *Treponema* (*T.*) *denticola* ATCC 35405 on ERK, p38 and JNK MAP kinases using non-keratinizing periodontal ligament epithelial cells (PLE) in vitro and discovered that compared to Chinese hamster ovary cells, human cervix adenocarcinoma cells, human osteosarcoma cells and human gingival fibroblasts, PLE cells were more resistant to *T. denticola*-induced reduction in cell viability. A low dose of 5×10^7 *T. denticola* cells/ml increased DNA synthesis in PLE cells but at higher concentrations DNA synthesis was decreased.

About 50% of epithelial cells in monolayers died through apoptosis when exposed to a high dose of 10^{11} *T. denticola*/ml for 24 h. Rounding, membrane damage, fragmentation and detachment were observed in selective cells of both mono- and multilayered PLE cultures exposed to *T. denticola*. In addition, *T. denticola* strongly but transiently activated ERK1, ERK2, JNK and p38 kinase. A specific inhibitor of the ERK MAP kinase pathway prevented the *T. denticola* stimulation of cell proliferation, whereas inhibitor of p38 elevated the cell numbers in *T. denticola*-treated cultures. So, *T. denticola* activated epithelial cell MAP kinase signal pathways responsible for cell proliferation and cell survival and exerted cytotoxic effects that appear to predominate at higher bacterial concentrations. Takeyama et al. (2006) revealed that membrane fraction of *M. hominis* induced IL-8 secretion from PC-3 cells and NF-κB activation in a concentration-dependent manner. Antibody against the extracellular TLR2 domain significantly suppressed *M. hominis*-induced IL-8 secretion from the prostate cell lines including PC-3, PrEC, and transformed myofibroblasts.

The small number of studies carried out does not give a definite conclusion, but the feasible connection between *M. hominis*, *Ureaplasma urealyticum* and prostate cancer certainly should attract attention to the problem. The correlation between the bacteria described above and gynecological cancer types seems to be elusive, although *G. vaginalis* and *Ureaplasma urealyticum* may possibly play a role in their etiology. So, it is possible to place these three infectious agents into the extended list of the infectious agents that may be a cause of cancer development but are not included into the IARC roster. Further epidemiological studies and research of fundamental carcinogenic properties of *M. hominis*, *Ureaplasma urealyticum* and *G. vaginalis* are needed for their establishment as carcinogenic agents of the short list. Population studies of the *M. hominis* significance in prostate cancer etiology should be performed not only in Russian population but in European, Asian and American as well. Although the results of existing studies are intriguing, they certainly should be replicated by other laboratories. For others, additional basic investigations on cell lines and animal models as well as population studies are required to consider them as possible etiological agents of cancer.

3.3.2 *Propionibacterium acnes*

Propionibacterium acnes (*P. acnes*) is suspected as a cause of several inflammatory conditions including acne vulgaris, endocarditis, endopthalmitis and osteomyelitis. It is a common microorganism colonizing human skin, but additional habitats include the oral cavity, large intestine and possibly the female genital tract. The history of investigation of association between *P. acnes* and cancer was started by Sheehan-Dare et al. (1988) who interviewed 70 patients with lymphomas and leukaemias. According to their data, acne was less common in the patients than in matched controls, suggesting a reduced incidence of acne or its more severe forms in patients who have developed lymphomas and leukemias. The authors hypothesized

Table 3.5 The connection between *P. acnes* and prostate cancer in epidemiological studies

Authors, year, population	Sample size	Methods of detection	Share of positive individuals, OR, 95% CI
Giles et al. (2003), Australian population	1,476 cases, 1,409 healthy controls	Self-reporting	0.67 (0.45–1.00)
Lightfoot et al. (2004), Canadian population	760 cases, 1,632 healthy controls	Self-reporting	No connection between acne and prostate cancer
Alexeyev et al. (2006), Swedish population	171 cases, 352 controls with BPH, 181 healthy controls	PCR	23% (cases) 2.17 (0.77–6.95)
Sutcliffe et al. (2007a, b), U.S. population	2,147 cases, 32,482 healthy controls	Self-reporting	1.70 (1.03–2.80)
Severi et al. (2010), Australian population	809 cases, 584 healthy controls	ELISA	50% (cases), 57% (controls) 0.73 (0.58–0.91)

that increased exposure or response to *P. acnes* may confer protection against malignant disease. Fifteen years later, Giles et al. (2003) investigated an association between prostate cancer risk and acne in a population-based, case-control study in Australia which included 1,476 cases and 1,409 controls in the period from 1994 to 1998. In this study, facial acne scarring non-significantly correlated with lower risk of prostate cancer in later life (OR = 0.67, 95% CI = 0.45–1.00). Lightfoot et al. (2004) carried out a population-based study in northeastern Ontario from 1995 to 1999, recruiting 760 prostate cancer registry-identified cases and 1,632 age-frequency matched controls, and found that acne were not significantly related to prostate cancer risk. Cohen et al. (2005) collected specimens of prostatic tissue from 34 consecutive Australian patients with prostate cancer. *P. acnes* was found in 35% of samples, and *P. acnes*-positive cases were characterized by a significantly higher degree of prostatic inflammation. As a consequence, this inflammation may be linked to the evolution of carcinoma. Moreover, organisms infecting these patients with prostate cancer differed genetically and phenotypically from the commonly identified cutaneous *P. acnes* isolates. Galobardes et al. (2005) reported a non-significant correlation between a history of acne and higher risk of prostate cancer mortality (hazard ratio = 1.67, 95% CI = 0.79–3.55) (Table 3.5).

Alexeyev et al. (2006) evaluated the presence of bacterial 16S RNA in archival prostate samples from 352 Swedish patients with benign prostate hyperplasia, 171 prostate cancer cases and 181 matched controls without cancer. 16S RNA was detected in 96/352 (27%) of the prostate tissue specimens, *P. acnes* was the predominant microorganism (23% of 16S RNA positive patients), and the presence of *P. acnes* in the prostate was non-significantly associated with prostate cancer risk (OR = 2.17, 95% CI = 0.77–6.95). Shannon et al. (2006) detected that *P. acnes* types IB and II were no more prevalent in the urinary tract of patients with prostate pathology than in healthy controls. A year later, Sutcliffe et al. (2007a, b) investigated severe acne, as measured by tetracycline use for 4 or more years, in relation to incident prostate cancer in the Health Professionals Follow-up Study. Between 1992 and 2002, 2,147 prostate cancer cases were observed among 34,629 eligible participants, and men who used tetracycline for 4 or more years had a significantly higher risk of prostate cancer (16 cases, 1,569 person-years) than men who did not use tetracycline (2,071 cases, 304,822 person-years, multivariable-adjusted RR = 1.70, 95% CI = 1.03–2.80). Shannon et al. (2008) investigated whether the serum titers of *P. acnes* antibodies in 68 patients undergoing prostate biopsy are associated with prostate cancer or markers of prostate disease. In patients with biopsies negative for cancer, a high anti-*P. acnes* antibody titer was associated with high serum PSA levels and was the predominant independent predictor of serum PSA level. The titer was positively correlated with patient age, prostate volume and aggressive inflammation, suggesting an involvement with benign prostatic hyperplasia (BPH). Severi et al. (2010) measured plasma concentration of *P. acnes* antibodies for 809 Australian cases and 584 controls using ELISA assay, and found that OR for prostate cancer associated with titers at or above the median value (1:1,024) was 0.73 (95% CI = 0.58–0.91). In addition, the correlation appeared to be particularly strong for advanced prostate cancer (OR = 0.59, 95% CI = 0.43–0.81). Last but not the least, the last study devoted to the problem was carried out by Fassi Fehri et al. (2011) who detected *P. acnes* by in situ immunofluorescence (ISIF) in 58 out of 71 (81.7%) tested cancerous prostate tissue samples, but not in healthy prostate tissues (20 samples) and other cancerous tissue biopsies (59 breast cancer samples). Active production of cytokines and chemokines (IL-6 and IL-8) from infected cells was observed, the host cell response was mediated by the transcriptional factors NF-κB and STAT3 activated upon *P. acnes* infection, and it was also mediated by activation of the COX2-prostaglandin and the plasminogen-matrix metalloproteinase pathways. Long-term exposure to *P. acnes* altered cell proliferation and caused anchorage-independent growth of infected epithelial cells.

Although the obtained data about the role of *P. acnes* in the etiology of prostate cancer are not consistent, it is feasible to place it into the extended list of the infectious agents that may be a cause of cancer development but are not placed in the IARC roster. If future epidemiological studies will find an association of *P. acnes* chronic carriage with increased prostate cancer risk, and new fundamental carcinogenic properties will be identified in basic research on cell lines and animal models, it will be possible to replace this bacterium in the short list.

3.3.3 *Neisseria gonorrhoeae and Treponema pallidum*

Neisseria gonorrhoeae (*N. gonorrhoeae*) is an exclusive pathogen of humans that infects the mucosal epithelia of the male urethra and the female lower genital tract. It is the etiological agent of gonorrhea, and, for women in particular, infection can spread into the upper reproductive tract, including the Fallopian tubes. In 1978, Furgyik and Astedt (1980) detected a 4-fold risk of cervical cancer in women who previously had gonorrhea on a sample of 164 German cases and the same number of controls.

They published the results as an original research paper in 1980. In 1983, Mommsen and Sell (1983) collected the data from 165 bladder cases and 165 controls and revealed that patients with a gonorrhea in the past were characterized by increased bladder cancer risk (RR = 2.42). The association of gonorrhea with cervical cancer was also reported in 1985 in Denmark by Lynge and Jensen (1985) and in Germany by Furgyik (1985). Two years later, Daling et al. were the first who tried to assess the role of *N. gonorrhoeae* infection as the possible risk factor for anal cancer. They interviewed 148 U.S. persons with anal cancer and 166 subjects with colon cancer and found that anal cancer was associated with a history of gonorrhea in heterosexual men (RR = 17.2, 95% CI = 2.0–149.4). In the same year, Furgyik (1987) in Germany found that women with a history of gonorrhea have 4.4-fold elevated cervical cancer risk compared to healthy controls. Herrero et al. (1990) recruited 759 cervical cancer patients and 1,430 controls in Latin America and demonstrated an increased risk of cervical cancer in women with gonorrheal history. An association between the gonorrhea and vulvar cancer risk (OR = 2.2 for in situ cancer and 5.0 for invasive carcinoma) was discovered on a 223 U.S. vulvar cancer cases and 466 healthy controls by Sherman et al. (1991). The positive connection of gonorrhea with bladder cancer was reported by La Vecchia et al. (1991) in Italy, who investigated 364 cases and 447 controls and obtained RR = 2.8 (95% CI = 1.0–4.5). Kjaer et al. (1992) identified the positive correlation between the history of gonorrhea and cervical cancer risk on 645 Danish cases and 614 controls. On the contrary, Duncan et al. (1993) did not find a correlation between gonorrhea and cervical cancer in Ethiopia. In 1994, de Sanjosé et al. (1994) conducted two case-control studies devoted to the same problem. In the first study, they collected 525 Spanish and Colombian CIN-III cases and 512 age- and place-matched controls. The sample from the second study included 373 histologically confirmed cases of squamous-cell cervical carcinoma and 387 age- and place-matched controls. Among the Spanish women, an elevation of invasive carcinoma risk was found for *N. gonorrhoeae*-positive subjects, but this association was valid only in HPV-DNA-negative women. Johansen et al. (2001) recruited a cohort of 4,440 women hospitalized for gonorrhea, and observed a total of 227 cases of CIN III with 103 expected (SIR = 2.2; 95% CI = 1.9–2.5), with no significantly increased risk for other anogenital cancers or cancer at other sites was seen. Williams et al. (1994) showed a positive association of gonorrhea with invasive cervical cancer (OR = 3.2, 95% CI = 1.6–6.2) on a Kenyan sample that included 112 cervical cancer cases and 749 controls. These results were further proved on a German sample of 309 cases with CIN or invasive cervical cancer and 490 controls by Köhler and Wuttke (1994).

Dennis et al. (2002) carried out a meta-analysis of investigations analyzing the connection between the prostatitis and prostate cancer, finding the overall OR = 1.6. Taylor et al. (2005) conducted a meta-analysis of 29 previous case-control studies devoted to the relation of sexually transmitted diseases (STDs) and prostate cancer, calculating the final OR = 1.48 (95% CI = 1.26–1.73) for STDs overall, OR = 1.35 (95% CI = 1.05–1.83) for gonorrhea, and OR = 1.61 (95% CI = 1.08–2.39) for syphilis. According to the noticed paper, the relation of prostate cancer to gonorrhea was investigated in 11 studies, to syphilis in 7 studies (Heshmat et al. 1975; Baker et al. 1981;

Lees et al. 1985; Mishina et al. 1985; Checkoway et al. 1987; Honda et al. 1988; La Vecchia et al. 1993; Hiatt et al. 1994; Ilić et al. 1996; Hsieh et al. 1999a, b; Hayes et al. 2000; Rosenblatt et al. 2001). After this meta-analysis, Pelucchi et al. (2006) performed a hospital-based case-control study in Italy, including 280 cases of prostatic cancer and 689 controls, and revealed that the OR of prostate cancer was 0.64 for subjects with history of selected venereal diseases and 0.53 for patients for prostatitis. So, the authors did not find any association between genital and urinary tract diseases and prostate cancer risk. Sarma et al. (2006) collected the data from 129 cases with history of gonorrhea and prostatitis and 703 controls, detecting the prostate cancer OR = 1.78 (95% CI = 1.13–2.79) and 4.93 (95% CI = 2.79–8.74) for gonorrhea and prostatitis, respectively. Sutcliffe et al. (2006) utilized data from 36,033 from the Health Professionals Follow-up Study and, on the contrary to previous results, observed no association between gonorrhea (RR = 1.04, 95% CI = 0.79–1.36), syphilis (RR = 1.06, 95% CI = 0.44–2.59), clinical prostatitis (RR = 1.08, 95% CI = 0.96–1.20) and prostate cancer, although a significant positive association was observed among men with clinical prostatitis younger than 59 years and prostate cancer (RR = 1.49, 95% CI = 1.08–2.06). Interestingly, Michaud et al. (2007) reported an increased bladder cancer risk among men with gonorrhea history (RR = 1.92, 95% CI = 1.10–3.33) utilizing the same data. Huang et al. (2008) did not find the association between gonorrhea, syphilis and prostate cancer risk in 868 cases (765 Whites and 103 Blacks) and 1,283 controls using the method of self-reporting. Löfmark et al. (2011) reported that upon *N. gonorrhoeae*-infection, a massive upregulation of amphiregulin mRNA is seen. The protein changed its subcellular distribution and was also alternatively cleaved at the plasma membrane that resulted in increased release of an infection-specific 36 kDa amphiregulin product from the surface of human cervical epithelial cells. Amphiregulin is a well-studied growth factor with functions in various cell processes and is upregulated in different forms cancer and proliferative diseases.

Syphilis is another sexually transmitted disease (STD) caused by *Treponema pallidum* (*T. pallidum*) subspecies pallidum, with an estimated 12 million new cases per year. Sherman et al. (1991) did not detect a connection of syphilis with vulvar cancer when the sample of 223 U.S. vulvar cancer cases and 466 healthy controls were recruited, and La Vecchia et al. (1991) could not reveal a correlation between syphilis and bladder cancer in Italy, investigating 364 cases and 447 controls. Köhler and Wuttke (1994) detected a positive association between syphilis and cervical cancer on a German sample of 309 cases with CIN or invasive cervical cancer and 490 controls. Michalek et al. (1994) conducted a prospective study, collecting sample of 16,420 patients with syphilis and analyzing the correlation with various cancer types. They found that that the number of patients suffering from malignant tumors of the oral cavity increased significantly in individuals who had syphilis (SIR = 169, 95% CI = 109–249), especially for cancer of the tongue (SIR = 251, 95% CI = 108–494). There was also a significant increase among males for Kaposi's sarcoma (SIR = 2,000, 95% CI = 1,290–2,950). In the study of Stone et al. (1995) who recruited 564 cervical cancer cases and 764 controls in Costa Rica, syphilis also was a risk factor of cervical cancer. Frisch et al. (1996) analyzed 1,523 Danish patients with penis cancer and did not observe a statistically significant association between syphilis and etiology of this disease. Tseng et al. (2003) interviewed 208 Taiwan anal cancer cases and 208 controls, indicating the syphilis as possible risk factor of anal cancer.

As there are no fundamental research analyzing the carcinogenic features of *N. gonorrhoeae* and *Treponema pallidum*, it is not possible to include them in the short list of the infectious agents that may be a cause of cancer development but are not placed in the IARC roster. However, it is feasible to place them in the extended list, and consistent results of extensive epidemiological studies make them an interesting target in investigating their oncogenic features on cell lines and animal models.

3.3.4 *Chlamydia trachomatis*

Primary infection with *C. trachomatis*, the most prevalent sexually transmitted bacterium worldwide with an estimated 90 million new cases occurring each year, is often asymptomatic and may persist for several months or years. The first study analyzing possible association of *C. trachomatis* with cervical cancer was carried out by Schachter et al. (1975) who assessed the prevalence of antibodies to TRIC (trachoma-inclusion conjunctivitis) agents in women with cervical dysplasia and in women attending selected clinics (obstetrics, cancer-screening, and gynecology). According to this investigation, antibodies to chlamydiae were identified in 77.6% of the women with dysplasia or cervical cancer whereas antichlamydial antibodies were less prevalent in the other clinic populations. Four years later, Paavonen et al. (1979) obtained the similar results in 93 of patients with cervical dysplasia comparing them to the controls. Furthermore, Cevenini et al. (1981) found a significantly higher incidence of histological cervical dysplasia in women with glandular ectopia (160 cases) and antichlamydial antibodies in comparison with *Chlamydia*-negative women. In the same year, Kalimo et al. (1981) observed IgA antibodies to *C. trachomatis*in cervical secretions of 24 of 35 (69%) patients with malignant or premalignant alterations, in 11 of 28 (39%) with cervicitis and in 3 of 26 (12%) controls, but serum IgA antibodies to *C. trachomatis* were found equally on the patients and the controls. Hare et al. (1982) isolated *C. trachomatis* in a significantly greater proportion of women with cervical cancer (18%) or CIN (8%) than in the control group (1%). In addition, histological examination of the cervix from 26 women with CIN and from 39 women of similar age without CIN revealed chlamydia-associated lymphoid follicles in 9 of the specimens with CIN, but in none of the CIN-negative specimens. In the second study of Schachter et al. (1982) in which they tested 383 women with and 500 women without cervical neoplasia for antibodies against *C. trachomatis*, a significant excess of antibodies against this bacterium was observed in cases (76.5%) compared to controls (58.4%). In the study of Cardillo (1985), 41.79% of smears from patients with CIN were associated which *C. trachomatis*-infection. Qvigstad et al. (1985) showed that the patients with CIN and invasive cancer are characterized by the similar degree of immune responses to chlamydial antigens (measuring levels of antibody titer and T-cell proliferative response). Guaschino et al. (1988) found that in patients with CIN 3 the rate of anti-*C. trachomatis* IgG was significantly higher (66.67% vs 11.76%) in comparison to the whole group (subjects with CIN or viral condylomata of the uterine cervix). Alberico et al. (1989) confirmed a greater frequency of anti-Chlamydia antibodies in cases with CIN 3 (75%). In the extensive investigation of Koutsky et al. (1992) who studied prospectively a cohort of 241 women, the presence of serum antibodies to Chlamydia trachomatis was associated with the development of CIN (RR=2.4, 95% CI=1.0–5.7). Hakama et al. (1993) analyzed 32 cervical cancer individuals and 64 matched controls in their prospective study and detected that after adjustment for smoking and other sexually transmitted diseases, antibodies to *C. trachomatis* showed the strongest association with cervical cancer (OR=5.0, 95% CI=1.6–15.7). Jha et al. (1993) collected serum samples from 219 women with cervical cancer and from 387 controls for anti-*Chlamydia* antibodies, revealing that *C. trachomatis*-infection may be associated with cervical cancer. However, in the case where HPV-seropositive cases and controls were compared, this association was weaker and non-significant. Muñoz et al. (1993) carried out a case-control study of 525 histologically confirmed cases of CIN III and 512 controls in Spain and Colombia, in which they found antibodies to *C. trachomatis* being associated with higher risk of CIN III (in Spain, OR=2.3, 95% CI=1.1–4.5; in Colombia, OR=1.7, 95% CI=1.1–2.7). In addition, the risk of CIN III was increasing with elevation of *C. trachomatis* antibody titers. Carta et al. (1994) obtained serologic and cytologic evidence of elevated exposure to *Chlamydia* among patients with squamous intraepithelial lesions. Dillner et al. (1994) collected sera from 94 cervical cancer cases matched against 188 age- and sex-matched controls in Northern Sweden and observed an increased risk of this malignancy for anti-*C. trachomatis* IgG-positive subjects (RR=1.7, 95% CI=1.0–2.7). A study carried out in Bulgaria by Borisov and Maĭnkhard (1995) in 128 women with cervicitis suggested genital chlamydial infection does not influence directly the CIN development, but these results were confronted even in the same year by Alaniz Sánchez et al. (1995) and González Sánchez et al. (1995) in Mexico. Hu and Gu (1995) in China found *C. trachomatis*-infection in a significantly greater proportion in women (in total 99 subjects) with cervical carcinoma (61.5%) and CIN (83.3%) compared to the control group (39.5%). Bosch et al. (1996) evaluated cytologic samples of 183 husbands of women suffering from invasive cervical cancer and of 171 husbands of control women in Spain which revealed that husband-positivity for antibodies to *C. trachomatis* was associated with cervical carcinoma in wife (adjusted OR=2.6, 95% CI=1.4–4.6). Their results were also confirmed in analogical study in Colombia by Muñoz et al. (1996) who obtained adjusted OR=2.5, 95% CI=1.5–4.4. In the same year, Lehtinen et al. (1996) detected a significant association between *C. trachomatis* carriage (particularly G, F, K serovars) and cervical cancer risk (adjusted OR=1.8, 95% CI=0.9–3.6 and adjusted OR=3.3, 95% CI=0.9–12.1, respectively) in a case-control study that included 72 cases and 143 matched controls in Finland. Additional evidence about the positive association between *C. trachomatis* and cervical cancer were provided by Dillner et al. (1997) who prospectively analyzed 182 invasive cervical cancer cases and 538 matched controls in Sweden, Norway and Finland, and found an increased risk of this tumor for *C. trachomatis* carriers (adjusted RR = 1.8, 95% CI = 1.2 = 2.9). On the contrary, Ferrera et al. (1997) could not observe such connection in their case-control study in Honduras (92 cases, 177 controls) and Takac and Gorisek (1998, 1999) did not find statistically significant association of *C. trachomatis* prevalence and various CIN stages as well as of *C. trachomatis* prevalence in patients with CIN and in control group. Nevertheless, Hsieh et al. (1999a, b) observed a significant positive correlation of *C. trachomatis* serotiter with cervical cancer risk in Taiwan (183 cases, 293 healthy controls, OR = 7.0, 95% CI = 1.7–28.9). The following year, Hakama et al. (2000) conducted a nested case-control study including 208 patients with invasive squamous cell cervical cancer and 624 matched controls, and detected that OR = 3.4 for anti-*C. trachomatis* IgG antibodies carriers. Their results were further confirmed by Koskela et al. (2000) and Lehtinen et al. (2002) in the same countries who obtained similar positive association (HPV- and smoking-adjusted OR = 2.2, 95% CI = 1.3–3.5). The association remained after adjustment for smoking both in HPV16-seronegative and -seropositive cases (OR = 3.0, 95% CI = 1.8–5.1 and OR = 2.3, 95% CI = 0.8–7.0, respectively). Anttila et al. (2001) investigated whether this risk is serotype-specific (on the same sample) and revealed that serotype G was most strongly associated with invasive squamous cell cervical carcinoma (adjusted OR = 6.6, 95% CI = 1.6–27.0), but other serotypes also had significance from this point of view, for instance, I serotype (OR = 3.8, 95% CI = 1.3–11.0) and D serotype (OR = 2.7, 95% CI = 1.3–5.6). The presence of serum IgG antibodies to more than 1 serotype increased the adjusted ORs ($P < 0.001$ for trend). Reesink-Peters et al. (2001) noted higher frequency of antichlamydial antibodies in patients with invasive cervical carcinoma (93%) than in women with CIN (35%), although they did not find significant differences between cervical cancer cases and control group in a second group of women. Smith et al. (2001, 2002) examined 499 women with incident invasive cervical cancer cases and 539 control patients from Brazil and the Philippines, detecting that *C. trachomatis* increased risk of squamous cervical cancer among HPV-positive women (OR = 2.1; 95% CI = 1.1–4.0). The results were similar in both countries. Wallin et al. (2002) identified 118 women with cervical cancer and selected 118 matched controls in Northern Sweden, detecting the adjustment for concomitant HPV DNA positivity relative risk for cervical cancer associated with previous *C. trachomatis* infection of 17.1-fold (95% CI = 2.6-infinity). Boyle et al. (2003) conducted a prospective, case-controlled, cross-sectional study and found that women with bacterial vaginosos did not have CIN more frequently than women with normal vaginal flora, and the quantities of nitrosamines produced by women with bacterial vaginosis did not differ significantly from women without BV. Castle et al. (2003) revealed that C. trachomatis is not associated with grade of cervical neoplasia when the presence of bacterial DNA and anti-*C. trachomatis* antibodies in 447 patients with distinct grade of CIN was assessed. Paavonen et al. (2003) recruited 178 women with invasive cervical carcinoma and 534 controls, founding antibodies to chlamydial heat shock protein 60-1 being associated with cervical squamous cell carcinoma among cases with long lag time (>3.5 years; OR = 2.4, 95% CI = 1.1–5.1). However, antibodies to chlamydial heat shock protein 60-2 or chlamydial heat shock protein 60-3 were not associated with cervical cancer risk. Hinkula et al. (2004) created the study sample of grand multiparous women (at least 5 children) including 220 cervical cancer cases and 178 CIN3 cases, assessing seroprevalence of *C. trachomatis* amongst them. They observed that the incidence of *C. trachomatis-infection* among grand multiparous women was slightly above the national average for squamous cell carcinoma of cervix uteri (SIR = 1.21, 95% CI = 1.05–1.40) and CIN3 (SIR = 1.37, 95% CI = 1.17–1.58), but lower for adenocarcinoma (SIR = 0.77, 95% CI = 0.52–1.10). Smith et al. (2004) carried out a case-control study of 1,238 cases of invasive cervical cancer and 1,100 control women from 7 countries (Thailand, Philippines, Morocco, Peru, Brazil, Colombia and Spain). Among HPV DNA-positive cases and controls, the risk of squamous cell invasive cervical cancer was higher in *C. trachomatis*-seropositive women (OR = 1.8, 95% CI = 1.2–2.7) after adjustment for age, center, oral contraceptive use, history of Pap smears, number of full-term pregnancies and herpes simplex virus 2 seropositivity. The effect of *C. trachomatis*-seropositivity on risk of this malignancy increased with elevating *C. trachomatis* antibody titers and was higher in women under 55 years of age. Nevertheless, *C. trachomatis* antibodies were not associated with adeno- or adenosquamous cell carcinoma (OR = 1.0, 95% CI = 0.53–1.9) in HPV DNA-positive women. In any case, the association of *C. trachomatis* with squamous cell invasive cervical cancer was found among all cases and control women with or without adjustment for HPV. Madeleine et al. (2007) investigated the *C. trachomatis*-cervical cancer association by serotype, histology, HPV type in the tumor, and other HPV cofactors, conducting a population-based study in the Seattle-Puget Sound area of 302 women with invasive cervical squamous cell carcinomas, 185 women with adenocarcinomas of the cervix, and 318 HPV-seropositive control women. They showed that the risk of invasive cervical squamous cell carcinomas associated with antibodies to *C. trachomatis* was increased (OR = 1.6, 95% CI = 1.1–2.2) but it was not true for adenocarcinoma (OR = 1.0, 95% CI = 0.6–1.5). This association was independent of HPV type in the SCC tumor tissue. In addition, there was an association between specific serotypes of *C. trachomatis* and SCC for certain serotypes: B (OR = 3.6, 95% CI = 1.5–8.4), D (OR = 2.1, 95% CI = 1.2–3.5), E (OR = 2.4, 95% CI = 1.4–3.9), G (OR = 3.0, 95% CI = 1.1–7.9), I (OR = 4.2, 95% CI = 1.5–11.7), and J (OR = 2.3, 95% CI = 1.0–5.1). Bułhak-Kozioł et al. (2007) recruited 75 patients with cervicitis accompanied by erosions, 45 patients with cervicitis alone, and 35 controls. *C. trachomatis* infection was found by direct immunofluorescence tests in 12.2% of patients with cervicitis alone with erosions, in 20% of patients with cervicitis alone, and in 2.9% of controls. IgG specific antibodies were detected in patients with cervicitis with erosions (34.7%), in 27.8% of patients with cervicitis and in 5.7% of controls. So, there was higher prevalence of *C. trachomatis* infection in female patients with cervical lesions as compared to unaffected women. In the prospective study of Naucler et al. (2007), in which they recruited a sample of 114 Taiwanese cervical cancer cases and of 519 matched controls, *C. trachomatis* carriage was not associated with cervical cancer risk, but it was associated with cervical cancer in cases restricted to cases identified within a follow-up period (OR = 2.94, 95% CI = 1.17–7.42) and in cases in which serum samples were analysed (OR = 3.13, 95% CI = 1.16–8.47). Oliveira Mde et al. (2008) conducted a cross-sectional study including 35 Brazilian patients with cytological cervical lesions and 35 control women and found the frequency of *C. trachomatis*-infection significantly higher in patients with CIN than in control group (80 and 14.3%, respectively, association prevalence ratio = 5.60, 95% CI = 2.44–12.82). Quint et al. (2009) collected biopsies of 71 Dutch patients with cervical adenocarcinomas using two novel PCR assays, but none of the cases was positive for *C. trachomatis*, suggesting that the carcinogenic role of this bacterium may be restricted to cervical squamous cell carcinomas. Safaeian et al. (2010) recruited 182 Costarican women with prevalent, 132 women with CIN2, CIN3and cervical cancer, and 995 controls. In this study, the authors found no association between *C. trachomatis* status (as assessed by DNA or IgG) at enrollment and combined prevalent and risk of cervical premalignancies and cancer. Valadan et al. (2010) collected data from 60 Iranic patients with CIN and 85 control subjects and revealed that antibodies to *C. trachomatis* were associated with an increased CIN risk (OR = 7.3, 95% CI = 1.5–35.2), whereas the presence of inclusion bodies for *C. trachomatis* and CIN (OR = 5.5, 95% CI = 2.4–12.4). Arnheim Dahlström et al. (2011) performed a large prospective study, retrieving serum samples from 604 Nordic women with invasive cervical cancer and from 2,980 matched controls and found previous exposure to *C. trachomatis* (was indicated by serum antibodies) being significantly associated with increased risk for cervical cancer (OR = 1.9, 95% CI = 1.5–2.3). According to the data of large prospective study including 8,441 women conducted by Lehtinen et al. (2011), the hazard ratios (HRs) associated with development of CIN2 to baseline *C. trachomatis* status were 1.82 (95% CI = 1.06–3.14) and 1.74 (95% CI = 1.05–2.90) (Table 3.6).

Ness et al. (2003) measured antibodies to *C. trachomatis* in 117 women with ovarian cancer and in 171 age- and ethnicity-matched population-based control subjects from Oahu, Hawaii. According to the authors' data, the probability of having ovarian cancer was 90% greater in women with the highest levels of anti-chlamydia antibodies compared with the lowest levels of anti-chlamydia antibodies, suggesting that past or chronic persistent infection with chlamydia may be a risk factor for ovarian cancer. In 2007, this hypothesis was further tested by Wong et al. (2007) who collected 41 serum samples of patients with ovarian cancer, 27 samples of patients with benign cystadenomas (22.1%) and 54 control samples (normal ovaries). Although there was a higher incidence of IgA seropositivity and lower incidence of IgG seropositivity in the ovarian cancer group compared with the other groups, it was not statistically significant. Five years after their first study, Ness et al. (2008) tested 521 U.S. women with ovarian cancer and 766 controls for IgG antibodies to serovar D of chlamydia elementary bodies (EBs), but, unexpectedly, the ovarian cancer odds ratio among women with the highest titers (>or = 0.40 OD units) was only 0.6 (95% CI = 0.4–0.9) that was contradictory to their earlier finding of elevated titers for anti-*C. trachomatis* antibodies amongst ovarian cancer patients. In addition, Idahl et al. (2010) analyzed samples of frozen ovarian tissues of 186 Swedish women with benign conditions, borderline tumors, and epithelial ovarian cancer as well as tissue from the contralateral ovary of 126 of these women for the

Table 3.6 The connection of *C. trachomatis* with cervical cancer in epidemiological studies

Authors, year, population	Sample size	Methods of detection	Share of positive individuals, OR, 95% CI
Kalimo et al. (1981)	35 cervical cancer and CIn cases, 28 controls with cervicitis, 26 healthy controls	IgA detection in cervical secretions	69% (cases), 39% (controls with cervicitis), 12% (healthy controls)
Hare et al. (1982)			18% (cases), 8% (CIN), 1% (controls)
Hakama et al. (1993), Finnish population	32 cases, 64 healthy controls	Serological tests	5.0 (1.6–15.7)
Jha et al. (1993)	219 cases, 387 healthy controls	Serological tests	A significant association (non-significant after adjustment for HPV persistence)
Dillner et al. (1994), Swedish population	94 cases, 188 healthy controls	ELISA	1.7 (1.0–2.7) IgG is the indicator
Hu and Gu (1995), Chinese population		Indirect immunoperoxidase assay	61.5% (cases), 83.3% (CIN), 39.5% (controls)
Lehtinen et al. (1996), Finnish population	72 cases, 143 healthy controls	ELISA, microimmunofluorescence (MIF)	For all serovars: 35% (cases), 21% (controls) 1.8 (0.9–3.6) For G, F, K serovars: 10% (cases), 4% (controls) 3.3 (0.9–12.1) IgG is the indicator
Dillner et al. (1997), Swedish, Norwegian, Finnish population	182 cases, 538 healthy controls	MIF	26% (cases), 14% (controls) 1.8 (1.2–2.9)
Hsieh et al. (1999a, b), Taiwan population	183 cases, 293 healthy controls	Serological tests	7.0 (1.7–28.9)
Hakama et al. (2000), Swedish, Norwegian, Finnish population	208 cases, 624 healthy controls	Serological tests	3.4 IgG is the indicator
Koskela et al. (2000), Swedish, Norwegian, Finnish population	182 cases, 538 healthy controls	MIF	26% (cases), 14% (controls) 3.0 (1.8–5.1) IgG is the indicator

(continued)

Table 3.6 (continued)

Authors, year, population	Sample size	Methods of detection	Share of positive individuals, OR, 95% CI
Anttila et al. (2001), Swedish, Norwegian, Finnish population	181 cases, 533 healthy controls	MIF	27% (cases), 13% (controls) 6.6 (1.6–27.0) IgG is the indicator
Lehtinen et al. (2002), Swedish, Norwegian, Finnish population	178 cases, 527 healthy controls	MIF	2.3 (0.8–7.0)
Smith et al. (2002), Brazilian, Philippinean population	499 cases, 539 controls	MIF	47.7% (squamous cell carcinoma cases), 29.6% (adenocarcinoma cases), 22.1% (controls) 2.1 (1.1–4.0) IgG is the indicator
Wallin et al. (2002), Swedish population	118 cases, 118 controls	PCR	8% (cases), 0% (controls) 17.1 (2.5–infinity)
Paavonen et al. (2003), Swedish, Norwegian, Finnish population	178 cases, 534 controls	Serological tests	2.4 (1.1–5.1) Antibodies to HSP 60-1 is the indicator
Smith et al. (2004), Thai, Philippinean, Morocco, Peruvian, Brazilian, Colombian, Spanish population	1,238 cases, 1,100 controls	MIF	53.2% (squamous cell carcinoma cases), 39.4% (adenocarcinoma cases), 30.8% (controls) 1.8 (1.2–2.7) IgG is the indicator
Madeleine et al. (2007), U.S. population	487 cases, 318 HPV-seropositive controls	ELISA, MIF	38.4% (squamous cell carcinoma cases), 28.1% (adenocarcinoma cases), 26.1% (controls) 1.6 (1.1–2.2)
Naucler et al. (2007), Taiwanese population	414 cases, 519 controls	MIF	36% (cases), 28.7% (controls) 3.13 (1.16–8.47) IgG is the indicator
Arnheim Dahlström et al. (2011), Nordic population	604 cases, 2,980 controls	Serological tests	1.9 (1.5–2.3)

presence of *C. trachomatis*, but all the tissue samples studied tested negative for this microorganism. In the following year, Idahl et al. (2011) published a study in which they collected plasma samples from 45 Swedish patients with ovarian cancer and 180 controls and analyzed the presence of anti-*C. trachomatis* IgG and IgA antibodies and chlamydial HSP60-1 IgG antibodies to *C. trachomatis*. There was no association of IgG and IgA antibodies with ovarian cancer, but chlamydial HSP60-1 IgG antibodies were significantly associated with type II ovarian cancer in women with plasma samples obtained >1 year prior to diagnosis. Shanmughapriya et al. (2012) analyzed tissue samples of 39 Indian patients with ovarian cancer by PCR, detecting that carriers of *C. trachomatis* are at significantly higher risk of development of this malignancy (OR=32; 95% CI=3.33–307.65) (Table 3.7).

Anttila et al. (2005) assessed the risk of exposure of prostate cancer to *C. trachomatis* by analyzing 738 cases of prostate cancer and 2,271 matched controls from Finland, Norway, and Sweden. They identified a statistically significant inverse association (OR=0.69, 95% CI=0.51–0.94) that was consistent by different serotypes. In addition, there was a consistent dose-response relationship. Thereby, it seems to be that *C. trachomatis*-infection does not elevate the risk of prostate cancer. Two years later, Sutcliffe et al. (2007a, b) also analyzed the possible association between plasma antibodies to *C. trachomatis* and prostate cancer risk, creating a study cohort of 691 U.S. cases and 691 matched controls, and found no association were observed between seropositivity to *C. trachomatis* and prostate cancer risk as well. Thence, Huang et al. (2008) examined the risk of prostate cancer in relation to serum antibodies to *C. trachomatis* in a prospective study which included 868 U.S. cases and 1,283 matched controls. For Afro-Americans, carriage of IgA antibodies to *C. trachomatis* was associated with elevated prostate cancer risk (OR=2.1, 95% CI=1.2–3.6). Dennis et al. (2009) analyzed 267 U.S. prostate cancer cases and 267 controls for antibodies against *C. trachomatis*, and found a statistically significant association (OR=1.8, 95% CI=0.96–3.38) for the antibodies in the serum collected more than 5 years before the reference date. Finally, in the recent study, Hrbacek et al. (2011) analyzed 329 prostate cancer patients and 105 controls with BPH, detecting that patients with prostate cancer were less often seropositive for *C. trachomatis* (OR=0.45, 95% CI=0.21–0.99) compared to men with BPH (Table 3.8).

The investigation done by Daling et al. (1987), who interviewed and obtained blood specimens from 148 patients with anal cancer, uncovered that this disease may be associated with seropositivity for *C. trachomatis* (RR=2.3, 95% CI=1.1–4.8) in women without a history of warts. Peters et al. (1995) revealed a possible association between increased IgG serum antibody to *C. trachomatis* and the risk of Caposi sarcoma. Bjørge et al. (1997) prospectively assessed 57 esophageal cancer cases and 171 matched controls in Norway for antibodies to specific microorganisms but did not find a significant association between *C. trachomatis*-seropositivity and risk of this malignancy. Anttila et al. (1998) discovered a possible connection between *C. trachomatis* and lymphoma risk on the Finnish sample of 72 cases and 72 controls in a seroepidemiological study, calculating overall lymphoma risk associated with the presence of *C. trachomatis* antibodies by ELISA of OR=2.4, 95% CI=1.1 5.0. Riska et al. (2006) conducted a retrospective seroepidemiological

Table 3.7 The connection of *C. trachomatis* with ovarian cancer in epidemiological studies

Authors, year, population	Sample size	Methods of detection	Share of positive individuals, OR, 95% CI
Ness et al. (2003), Hawaiian population	117 cases, 171 healthy controls	ELISA	1.9 among women with the highest compared with the lowest levels of anti-chlamydia antibodies IgG is the indicator
Wong et al. (2007)	41 cases, 27 controls with benign cystadenomas, 54 healthy controls	Serological tests	No significant association
Ness et al. (2008), U.S. population	521 cases, 766 healthy controls	ELISA	Among women in the highest quartile of chlamydia EB antibodies compared to the lowest: 0.6 (0.4–0.9) IgG is the indicator
Idahl et al. (2010), Swedish population	186 cases, 126 control tissues	PCR	No detection of microorganism
Idahl et al. (2011), Swedish population	45 cases, 180 controls	MIF	HSP60-1 IgG antibodies were significantly associated with type II ovarian cancer in women with plasma samples obtained >1 year prior to diagnosis
Shanmughapriya et al. (2012), Indian population		PCR	32 (3.33–307.65)

Table 3.8 The connection of *C. trachomatis* with prostate cancer in epidemiological studies

Authors, year, population	Sample size	Methods of detection	Share of positive individuals, OR, 95% CI
Anttila et al. (2005), Finnish, Norwegian, Swedish population	738 cases, 2,271 healthy controls	MIF	7.5% (cases), 10.5% (controls) 0.69 (0.51–0.94) IgG is the indicator
Sutcliffe et al. (2007a, b), U.S. population	691 cases, 691 healthy controls	ELISA	4% (cases), 3.5% (controls) 1.13 (0.65–1.96), non-significant IgG is the indicator
Huang et al. (2008), U.S. population	868 cases, 1,283 healthy controls	ELISA	For Afro-Americans: 2.1 (1.2–3.6) IgG is the indicator
Dennis et al. (2009), U.S. population	267 cases, 267 controls	MIF	For the antibodies in the serum collected more than 5 years before the reference date: 1.8 (0.96–3.38)
Hrbacek et al. (2011), Czech population	329 cases, 105 controls with BPH	ELISA	5.5% (cases), 11.4% (controls) 0.45 (0.21–0.99) IgG is the indicator

study investigating the connection between past chlamydial infection and primary fallopian tube carcinoma in 79 patients and 142 controls. Seropositivity in general or serum antibody levels to distinct *C. trachomatis* serovars did not differ between primary fallopian tube carcinoma patients and controls. According to the data of Chanudet et al. (2007), *Chlamidya* spp. (*C. trachomatis*, *C. pneumoniae*, *C. psittaci*) may also affect the risk of pulmonary MALT lymphoma since the frequency of their detection in tissue samples of this type of tumors was significantly higher than in patients with non-lymphoproliferative disorders (35 and 18%, respectively, 69 cases and 44 controls). Zhang et al. (2009) did not find the presence DNA of *C. trachomatis* in 38 Chinese cases of primary ocular adnexal MALT lymphoma, 3 cases of non-MALT lymphoma and 27 case of chronic inflammation. Olejek et al. (2009) analyzed the frequency of detecting *C. trachomatis* antigens and antibodies against this bacterium in 30 patients with vulvar cancer and 50 women with lichen sclerosus vulvae as a control group, identifying anti-*C. trachomatis* IgG antibodies in 50% of vulvar cancer patients and in 16% women with lichen sclerosus vulvae.

Heggie et al. (1986) in their first attempt evaluated *C. trachomatis* (serovar LGV-2) for oncogenicity in a mouse model. The frequencies of cervical abnormalities in mice exposed to *C. trachomatis* or corresponding control fluid were low, and differences between the two groups were not statistically significant, suggesting that the LGV-2 serovar of *C. trachomatis* is not potentially oncogenic. Markowska et al. (1999) revealed the elevated expression of TGF-α and Ki67 in various stages of CIN and in cervical cancer, noticing that *C. trachomatis*-infection frequently leads to cervical hypertrophy. In addition, *C. trachomatis* infection may increase risk of HPV-infection, according to the data of Giuliano et al. (2001). Fischer (2002) collected a study sample of 81 women with CIN and 68 controls and analyzed the influence of *C. trachomatis*-infection on expression of EGFR, TGF-alpha, Ki 67, HPV 16 and 18. The author detected that chlamydial infection may cause cervical hypertrophy in women with and without CIN and invasive carcinoma. *C. trachomatis*-infection correlated with high expression of EGFR, TGF-α and Ki 67 in epithelium and increased the expression of HPV16 in CIN I, suggesting that *C. trachomatis*-infection modifies the activity of viruses. Prozialeck et al. (2002) observed that *C. trachomatis* caused the epithelial cells to separate from each other without detaching from their growing surface. In noninfected cells, N-cadherin and beta-catenin were colocalized at the intercellular junctional complexes. By contrast, the infected cells showed a marked loss of both N-cadherin and beta-catenin labeling from the junctional complexes and the further appearance of intense beta-catenin labeling associated with the chlamydial inclusion, indicating that *C. trachomatis* causes the breakdown of the N-cadherin/beta-catenin complex and that the organism can sequester beta-catenin within the chlamydial inclusion.

According to the plethora of basic and epidemiological studies carried out, *C. trachomatis* may be established as an etiological agent of cervical cancer, but results are widely distinct from each other in different countries. For ovarian cancer, the results are contradictory but still suggest a feasible positive association of *C. trachomatis* and risk of this malignancy in some geographical regions (for instance, Sweden or India). In the case with prostate cancer, the data are even more discrepant but it seems to be that the connection of *C. trachomatis* and this cancer type may be true for Afro-Americans. Concerning other malignant tumors, it is possible that *C. trachomatis* affect the risk of various lymphoma subcategories and vulvar cancer. In addition, there are a number of fundamental investigations confirming the carcinogenicity of *C. trachomatis*. According to our criteria, it is possible to place *C. psittaci* in the short list of the infectious agents that may be a cause of cancer development but are not included in the IARC roster.

3.4 The Role of Bacteria in Cancer of the Hematolymphopoietic System

3.4.1 *Borrelia spp.*

Lyme disease, caused by the tick-borne bacteria *Borrelia burgdorferi* (*B. burgdorferi*), is the most commonly reported vector-borne infectious disease in North America. The number of yearly cases reported to the CDC has steadily increased since 1982 when case reporting began, with 20,000–30,000 cases now reported each year. The history of investigation of the association of *B. burgdorferi* with cancer started from the study of Garbe et al. (1988) who presented two cases of malignant B-cell lymphoma of the skin in patients with chronic *B. burgdorferi* infection and concluded that an elevated titer of antibodies and certain histological characteristics indicating *Borrelia* infection is an important finding for the diagnosis and prognosis of this particular type of cutaneous B-cell lymphoma. In the same year, Uldry et al. (1988) reported a case of a 50-year-old man who had a primary lymphoma of the central nervous system 1 year after a tick bite and a radiculoneuritis secondary to *B. burgdorferi* infection. Examination of antibodies against *B. burgdorferi* revealed elevated IgM and IgG titers, and the autopsy demonstrated a primary B-cell immunoblastic lymphoma of the brain without evidence of extraneural lymphoproliferative disorder. Based on the case, the authors suggested *B. burgdorferi* as a possible cofactor which may provoke cellular proliferation and the formation of a lymphoma. Three years later, Garbe et al. (1991) presented four more patients with low-grade malignant B cell lymphoma of the skin in association with chronic *B. burgdorferi* infection based on specific clinical and immunological signs of Borrelia infection. Two more new cases of low-grade malignant primary cutaneous B cell lymphoma in association with *B. burgdorferi* infection were reported by Kütting et al. (1997), and in the same year Cerroni et al. (1997) in Austria conducted the first epidemiological study devoted to the association of *B. burgdorferi* with cancer risk. They retrieved 50 cases of primary cutaneous B-cell lymphoma (PCBCL) and performed PCR analysis of *B. burgdorferi* DNA on paraffin-embedded tissue sections. Specific DNA sequences of *B. burgdorferi* were identified in cutaneous tumors from 9 (18%) patients, and specificity was confirmed by Southern blot hybridisation in all positive cases. The authors proposed that the proportion of

PCBCL may be cured with antibiotic therapy against *B. burgdorferi*, and declared that PCR analysis of *B. burgdorferi* DNA is a rapid test that should be performed in all patients with PCBCL to identify subjects who can likely benefit from an early antibiotic treatment. The next large epidemiological study was carried out in Serbia by Jelić and Filipović-Lješković (1999) who collected samples from 30 patients with PCBCL and tested 22 of them for *Borrelia* using serological methods (ELISA, Western Blot). The control group consisted of 85 patients with NHL (10 cutaneous T-cell, 25 extranodal B-cell non-PCBCL, 50 nodal B-cell), 30 patients with breast cancer and 60 blood donors. The prevalence of positive borrelial serology was 12 out of 22 cases (55%) in the PCBCL group. No positives were detected in the cutaneous T-cell lymphoma group; 2 of 25 patients (8%) were positive in the extranodal B-cell NHL group, 2 out of 50 (4%) were positive in the nodal B-cell NHL group, 2 of 30 (7%) in the breast cancer group and 2 of 60 (3%) in the blood donor group. The total prevalence in the control groups was 8 out of 175 (4,6%), thus the incidence was significantly higher in PCBCL patients compared to each of the control groups. Goodlad et al. (2000a) analyzed the presence of *B. burgdorferi* in 20 cases of PCBCL from the Scottish Highlands, an area with endemic Lyme disease, and compared their findings with those in 40 control patients (20 undergoing wide reexcision at sites of malignant melanoma and 20 biopsies of inflammatory dermatoses). A nested PCR was performed on DNA extracts from the lymphoma and control cases using primers to a unique conserved region of the *B. burgdorferi* flagellin gene. *B. burgdorferi*-specific DNA was detected in 7 of 20 lymphoma cases and in 1 patient of 40 control subjects. The relationship between *B. burgdorferi* and PCBCL was significant when compared with the control groups separately or in combination, providing strong evidence to support the concept of *B. burgdorferi*-driven lymphomagenesis in the skin. Furthermore, in three other studies published in the same year Goodlad et al. (2000b), Roggero et al. (2000) and Hatanaka et al. (2000) reported four more cases of *B. burgdorferi*-associated PCBCL. Reports of the presence of *B. burgdorferi* DNA in malignant lymphomas have raised the hypothesis that infection by this bacterium may be causally related to non-Hodgkin lymphoma (NHL) development. Schöllkopf et al. (2008) carried out a Danish-Swedish case-control study including 3,055 NHL patients and 3,187 controls. ELISA for antibodies against *B. burgdorferi* was conducted in a subset of 1,579 patients and 1,358 controls. Overall risk of NHL was not significantly associated with self-reported history of tick bite (OR=1.0; 95% CI=0.9–1.1), *Borrelia* infection (OR=1.3, 95% CI=0.96–1.8) or the presence of anti-*Borrelia* antibodies (OR=1.3, 95% CI=0.9–2.0). Nevertheless, in the analyses of NHL subtypes, self-reported history of *B. burgdorferi* infection (OR=2.5, 95% CI=1.2–5.1) and seropositivity for anti-*Borrelia* antibodies (OR=3.6, 95% CI=1.8–7.4) were both associated with risk of mantle cell lymphoma. Notably, this specific association was also observed in persons who did not recall *Borrelia* infection yet tested positive for anti-*Borrelia* antibodies (OR=4.2, 95% CI=2.0–8.9). Moreover, the number of other reports reporting cases of *B. burgdorferi*-associated lymphomas of leukemia cutaneous manifestations was published by Cerroni et al. (2002), Bogle et al. (2005), Monari et al. (2007), Fühler et al. (2010), Kash et al. (2011) (Table 3.9).

Table 3.9 The connection between *B. burgdorferi* and lymphoma in epidemiological studies

Authors, year, population	Sample size	Methods of detection	Share of positive individuals, OR, 95% CI
Jelić and Filipović-Ljesković (1999), Serbian population	22 PCBCL cases, 85 NHL controls, 30 breast cancer controls, 60 healthy controls	ELISA, Western blotting	55% (cases), 5% (NHL controls), 7% (breast cancer controls), 3% (healthy controls)
Goodlad et al. (2000a, b), Scottish population	20 PCBCL cases, 40 controls	PCR	35% (cases), 2.5% (controls)
Schöllkopf et al. (2008), Danish and Swedish population	3,055 NHL cases, 3,187 controls	ELISA, self-reporting	1.3 (0.9–2.0), non-significant For mantle cell lymphoma: 3.6 (1.8–7.4) Among persons with both self-reported and serologic evidence of Borrelia infection: 3.5 (1.3–9.4)

On the other hand, Munksgaard et al. (2000) performed a correlation analysis of PCBCL and Lyme disease using data from the Surveillance, Epidemiology and End Results program and from the Centers for Disease Control and Prevention, but, unexpectedly, were not successful to demonstrate a geographic correlation between incidence rates of Lyme disease and PCBCL. It was suspected by the authors that infection with *B. burgdorferi* is not a major risk factor for PCBCL in the USA. Similarly, Wood et al. (2001) once more performed a U.S. epidemiological study analyzing the association between *B. burgdorferi* and cancer. They retrieved 38 cases of primary and secondary CBCL, but detected no *B. burgdorferi*-specific DNA in any of the 38 CBCL cases, proposing that the geographic variations in the clinical manifestations of *B. burgdorferi* are real and may be caused by genetic and phenotypic differences between bacterial strains in Europe and North America. Analogically, Li et al. (2003) examined 24 Asian cases of cutaneous marginal zone B-cell lymphoma for *B. burgdorferi* involvement, but no *Borrelia* DNA was detected, suggesting that *B. burgdorferi* infection may not play an important role in developing cutaneous marginal zone B-cell lymphoma in Asian cases. Kodama et al. (2005) examined 56 samples from Austrian patients with large BCL (LBCL) but could not find any presence of *B. burgdorferi* infection in those cases. Similar results were obtained in the epidemiological study of Goteri et al. (2007) who investigated 73 Italian cases of PCBCL, in the study of Takino et al. (2008) who analyzed 60 PCBCL cases from East Asia, Germany, and the United States, and in the investigation of Ponzoni et al. (2011a, b) who collected 98 Italian patients with PCBCL. The last large population study devoted to the problem was carried out in Sweden by Chang et al. (2012) who followed-up 107,177 total patients (of these, 15,249 *Borrelia*-positive) for 6.6 years and revealed no increased risk of any cancer type.

According to the results of Böggemeyer et al. (1994), the preparations of the *B. burgdorferi* upregulated cell surface expression of E-selectin, P-selectin, ICAM-1 and VCAM-1 on a mouse endothelioma cell line (bEnd 3 cells) in a dose-dependent manner, suggested that E-selectin, P-selectin, ICAM-1 and VCAM-1 play an important role in the pathogenesis of spirochetal infection. Ebnet et al. (1996) showed that inflammatory cytokines and bacterial structures such as lipopolysaccharide (LPS) and preparations of *B. burgdorferi* induced rapid chemokine production (MCP-1, IL-8, RANTES, IP-10) in endothelial cells. *B. burgdorferi* antigens which upregulated six chemokine genes and genes encoding adhesion molecules in both endothelial cells and fibroblasts showed that chemokines might play a central role in the regulation of spirochete-induced immune response. According to data Ebnet et al. (1997) obtained in the following year, *B. burgdorferi* induced rapid production of chemokines RANTES (regulated upon activation, normal T cells expressed and secreted), MCP-1 (monocyte chemoattractant protein-1), IL-8, GRO-α, IP-10 (IFN-inducible protein-10), and MIG (monokine induced by gamma-IFN), and of the adhesion molecules E-selectin, ICAM-1, and VCAM-1 in endothelial cells and production of the same chemokines and ICAM-1 in fibroblasts. This effect was mediated by the lipid moiety of the outer surface lipoprotein A. Most importantly, this induction did not require new protein synthesis and could be blocked by inhibitors of NF-κB. In addition, *B. burgdorferi* induced nuclear translocation of NF-κB and a transient increase in the expression of the IκB-α. Therefore, this research group discovered that *B. burgdorferi* is a potent chemokine inducer, and the data suggested that this biologic activity is due to the ability of the spirochetes to activate the pleiotropic transcription factor NF-κB.

Obviously, the data obtained by different authors are contradictory, and it is hard to conclude whether *B. burgdorferi* is associated with lymphoma risk. The results of Asian and U.S. studies certainly do not support the hypothesis about the role of *B. burgdorferi* in lymphomagenesis, the results of European investigations are different even in the same countries (Austria, Sweden) and may reflect the diversity of bacterial strains, environmental conditions, features of populations and their subgroups etc. (in other words, host, environmental, and bacterial factors). New population studies in other countries may help to clarify the problem. Existing doubts in the prevalence of *B. burgdorferi* in various lymphoma types do not allow placing this bacterium in the short list of the infectious agents that may be a cause of cancer development but are not included in the IARC roster, but it is possible to include it to the extended list. In addition, basic research devoted to the problem is also insufficient and require further investigations of the carcinogenic properties of this infectious agent.

3.4.2 *Chlamydia psittaci*

Chlamydia psittaci (*C. psittaci*) is the etiologic agent of psittacosis, a human lung infection caused by exposure to infected birds. The occurrence of high rates of chlamydial infection in household cats and of asymptomatic carriage of *C. psittaci* in cats from breeding catteries raises the possibility that human *C. psittaci* infections derived from pets other than birds may be underdiagnosed. In 2004, Ferreri et al. (2004) in Italy first assessed whether *Chlamydiae* infection can be associated with cancer (namely, with the development of ocular adnexal lymphoma). They tried to identify *C. psittaci* DNA by PCR in 40 ocular adnexal lymphoma samples, 20 nonneoplastic orbital biopsies, 26 reactive lymphadenopathy samples, and in peripheral blood mononuclear cells (PBMCs) from 21 lymphoma patients and 38 healthy subjects, and found 32 out of 40 (80%) ocular adnexal lymphoma samples as *C. psittaci*-positive. On the contrary, none of the 20 nonneoplastic orbital biopsies and only 3 of 26 (12%) reactive lymphadenopathy samples (12%) carried the *C. psittaci* DNA. In addition, *C. psittaci* DNA was identified in PBMCs of 9 of 21 (43%) patients with *Chlamydia*-positive lymphomas whilst none of the healthy PBMC donors had *C. psittaci* DNA in their PBMCs. The authors concluded that patients with ocular adnexal lymphoma had a high prevalence of *C. psittaci*-infection in both tumor tissue and PBMCs in contrast to control tissues and PBMCs, and postulated that persistent *C. psittaci*-infection may contribute to the development of these lymphomas. This was also supported by the clinical response revealed in this study with *C. psittaci*-eradicating antibiotic therapy. In the following year, Ferreri et al. (2005) showed that after doxycycline assumption for 3 weeks, the objective response was complete in two *C. trachomatis*-positive patients with marginal-zone B-cell lymphoma of the ocular adnexa, partial response (>50%) was observed in two patients, and minimal response (<50%) was observed in three patients, proposing *C. psittaci*-eradicating antibiotic therapy with doxycycline as a new possible effective method of objective response achievement in patients with ocular adnexal lymphoma. A year later, Chanudet et al. (2006) assessed the presence of *C. psittaci* in 142 ocular adnexal MALT lymphomas, 53 non-marginal zone lymphomas, and 51 ocular adnexal biopsies without a lymphoproliferative disorder (LPD) from six geographical regions. In this study, *C. psittaci* was detected at similar low frequencies in non-LPD and non-marginal zone lymphoma groups from different geographical regions (0–14%); however, the prevalence of *C. psittaci* was significantly higher in MALT lymphomas (22%) than in non-LPD (10%) and non-marginal zone lymphoma cases (9%). The prevalence of *C. psittaci* infection in MALT lymphoma demonstrated a marked variation among all of the six geographical regions, being most frequent in Germany (47%), followed by the USA East Coast (35%) and the Netherlands (29%), but it was rather low in Italy (13%), the UK (12%), and Southern China (11%). In addition, Daibata et al. (2006) reported that they did not detect *C. psittaci* DNA in tumor tissues of 18 patients with MALT lymphomas, 3 non-MALT lymphomas of the ocular adnexa and 3 reactive ocular lymphoid hyperplasias by methods that were similar to those used in previous studies. However, Liu et al. (2006) detected *C. psittaci* DNA in three out of 12 Japanese patients with ocular adnexal MALT lymphoma by ethidium bromide staining of the PCR product, but not by Southern hybridization, showing that details of the detection method to identify *C. psittaci*-infection in ocular adnexal lymphoma may play a role in interpreting the study results. In addition, De Cremoux et al. (2006) assessed the frequency of detecting the *C. psittaci* DNA in biopsies of 16 French patients with ocular adnexal lymphoma,

8 patients with marginal zone B-cell lymphoma, 1 patient with splenic-type marginal zone B-cell lymphoma, 1 patient with follicular lymphoma, 10 patients with reactive lymphoid hyperplasia and from tissues of two control subjects with other types of lymphoproliferative disease. *C. psittaci* DNA was detected in the tumor tissue of only one patient with follicular ocular adnexal lymphoma that was much lower than in the study of Ferreri et al. (2004). Mulder et al. (2006) recruited a cohort of Dutch patients suffering from ocular adnexal marginal zone B cell lymphomas, but obtained no evidence of the presence of *C. psittaci* DNA in any of the tumor samples studied. Rosado et al. (2006) evaluated the association of primary ocular adnexal lymphoma with *C. psittaci*-infection in 57 U.S. patients (South Florida), but none of the specimens harbored C psittaci DNA. Vargas et al. (2006) analyzed a sample of patients from New York State with ocular adnexal lymphomas and benign lesions attempting to identify *C psittaci* DNA, but no evidence of its presence was seen in seven MALT-type ocular adnexal lymphomas, four non-MALT ocular lymphomas, one Langerhans histiocytosis, and five reactive lymphoproliferations. Chanudet et al. (2007) identified *C. psittaci* DNA only in 2 out of 35 (6%) of U.K. patients, in 1 of 21 German subjects and in 1 of 13 Italian patients with pulmonary MALT lymphoma but in none of 30 individuals with other lymphoproliferative disorders and in none of 44 subjects with non-LPD, although these differences did not reach statistical significance level. Gracia et al. (2007) obtained DNA samples from formalin-fixed, paraffin-embedded sections of 26 patients with ocular adnexal lymphomas, 20 subjects with other lymphomas and 20 benign ocular lesions and found the DNA of *C. psittaci* in 11% of the 46 lymphomas (in 2 orbital extranodal marginal zone B-cell lymphomas and 3 other lymphomas) whereas all 20 benign ocular lesions were negative for C. psittaci. Husain et al. (2007) carried out the meta-analysis devoted to the association between *C. psittaci* and ocular adnexal lymphoma across various geographic regions. The authors identified 11 studies which included 458 ocular adnexal lymphoma cases from 10 different countries, and calculated that 104 of the 458 OAL specimens (23%) and 87 of the 346 mucosa-associated lymphoid tissue (MALT) lymphoma specimens (25%) were *C. psittaci*-positive. Interestingly, 94 of these 104 cases (90%) came from 3 of the 11 studies, so wide variation between geographic regions and even between studies from the same geographic region were detected. In addition, four studies concerning the efficacy of antibiotics for ocular adnexal lymphoma treatment were found (Yeung et al. 2004; Abramson et al. 2005; Ferreri et al. 2005; Grünberger et al. 2006) including 42 patients. It was observed that 28 patients (48%) achieved a response (complete response in 8 patients, partial response in 8 patients, and minimal response in 4 patients), 20 patients had stable disease, and 2 patients progressed during antibiotic therapy, so the authors concluded that antibiotics have variable efficacy against OAL. Thereafter, Ruiz et al. (2007) investigated 34 U.S. patients with ocular adnexa MALT lymphomas, but could not detect *C. psittaci* DNA by PCR in all of these cases. Yakushijin et al. (2007) analyzed 23 ocular adnexal MALT lymphoma cases in the South Japan for the presence of *C. psittaci* DNA, but none of them was positive. In contrast, Yoo et al. (2007) recruited 33 Korean subjects with ocular adnexal lymphoma and detected *C. psittaci* DNA by PCR in 26 of 33 (79%) ocular adnexal lymphoma

samples compared with 5 out of 21 (23%) individuals with non-neoplastic ocular adnexal diseases which suggested that *C. psittaci* may play a role as a causative antigen in Korean patients with ocular adnexal lymphoma. However, Zhang et al. (2007) collected 28 ocular adnexal lymphoma biopsy specimens from adult U.S. patients, 16 control lymphoma specimens from patients with systemic lymphomas not involving the ocular adnexa, and 5 control benign adnexal tissue samples, but did not detect the presence of *C. psittaci* in ocular adnexal lymphomas. The following year, Aigelsreiter et al. (2008) identified *C. psittaci* DNA in 7 out of 13 Austrian ocular adnexal MALT lymphoma cases but only in 1 of the 17 control gastrointestinal specimens (from patients suffering from gastrointestinal MALT lymphoma and gastritis). Ferreri et al. (2008a) carried out a single-center prospective case-control study in Italy to assess the prevalence, viability and infectivity of *C. psittaci* in 20 patients with ocular adnexal MALT lymphoma and in 42 blood donors. *C. psittaci* DNA was detected by PCR in lymphoma tissue of 15 (75%) patients. Viability and infectivity of *C. psittaci*, indicated by growth in culture, were demonstrated in conjunctival swabs and/or PBMC from 5 (25%) patients, but not in donors. The authors concluded that *C. psittaci* present in the conjunctiva and PBMC of ocular adnexal MALT lymphoma patients can grow and be isolated in cell cultures. In the same year, this research group (Ferreri et al. 2008b) demonstrated the efficacy of doxycycline treatment (objective tumor response) in three out of six patients with ocular adnexal MALT lymphoma. Ponzoni et al. (2008) recruited 35 Italian ocular adnexal lymphoma patients, 205 subjects with extraorbital lymphoma and 135 nonneoplastic controls. The authors detected that 26 out of 25 (74%) of ocular adnexal lymphomas were associated with *C. psittaci*-infection using multiple methods: Immunohistochemistry, immunofluorescence, and laser-capture microdissection-assisted PCR, and showed that monocytes/macrophages were the *C. psittaci* carriers; in addition, the electron microscopy showed the presence of intact *C. psittaci* elementary bodies in these cells. However, Matthews et al. (2008) examined 49 patients with ocular adnexal MALT lymphomain South Florida for *C. psittaci* DNA but could not reveal it. Zhang et al. (2009, 2010) collected 68 archival cases of primary ocular adnexal lymphoproliferative lesions from Chinese patients, including 38 cases of MALT lymphoma, 3 cases of non-MALT lymphoma and 27 cases of chronic inflammation, but could not find *C. psittaci* DNA in these samples. Aigelsreiter et al. (2011) analyzed the presence of *C. psittaci* in 47 nongastrointestinal, 14 gastrointestinal MALT lymphomas, 37 nonmalignant control samples, and 27 autoimmune precursor lesions. In 47 nongastrointestinal MALT lymphomas, 13 (28%) tested positive for *C. psittaci* DNA compared with 4 (11%) of 37 nonmalignant control samples, but the results did not reach a statistical significance level, and only 1 (7%) of 14 gastrointestinal MALT lymphomas tested positive for *C. psittaci*. Finally, Ponzoni et al. (2011a, b) firstly reported a case of primary central nervous system marginal zone B-cell lymphoma involving the choroid plexus associated with *C. psittaci* infection and Cai et al. (2012) did not detect *C. psittaci* DNA in 16 tumor specimens of conjunctival MALT lymphoma from Chinese patients.

To sum up, there are basic as well as extensive epidemiological evidence that *C. psittaci* may be an etiological agent of ocular adnexal MALT lymphoma, but results are widely distinct from each other in different countries. For instance, in Italy, Austria and Korea the positive association of *C. psittaci* and this malignancy was stably detected, for other geographical regions no such connection was found. In addition, there is a lack of basic research devoted to the problem. According to our criteria, it is feasible to place *C. psittaci* in the extended list of the infectious agents that may be a cause of cancer development but are not included in the IARC roster.

3.5 Additional Bacteria Anecdotally Associated with Cancer

3.5.1 *Bartonella henselae, Tropheryma whippelii, Staphylococcus lugdunensis, Porphyromonas gingivalis*

There are only a few studies that have analyzed the possible role of Bartonella in cancer etiopathogenesis. In the study of Kirby (2004), *Bartonella* (*B.*) *henselae* induced endothelial proliferation, inhibition of apoptosis of endothelial cells, matrix invasion of endothelial cells, long-term endothelial survival and tubular differentiation within type I collagen-rich extracellular matrix, and increased VEGF synthesis from surrounding non-endothelial cells. These results partially proved the results of other studies carried out previously by Kempf et al. (2001), Resto-Ruiz et al. (2002), Kirby and Nekorchuk (2002), Dehio (2003). In addition, Gillen et al. (1993), Mohm et al. (1998), Gruner et al. (2001), Wang et al. (2003), and Bargiacchi et al. (2010) reported a series of cases of patients with various lymphomas feasibly associated with *Tropheryma whippelii* infection. Noguchi et al. (2007) revealed that tannase-producing *Staphylococcus* (*S.*) *lugdunensis* was significantly more frequently isolated from the patients with advanced-stage colorectal cancer (22.2%) than from the early-stage colorectal cancer cases (8.6%) or subjects with adenoma (4.9%).

There were a number of fundamental studies which analyzed the carcinogenic properties of *Porphyromonas* (*P.*) *gingivalis*. The studies found that this bacterium can attach itself to various gingival carcinoma cell lines (Hirose et al. 1996), invade gingival epithelial cells *in vitro* (Kim et al. 2010a, b), prevent and inhibit apoptosis of epithelial cells (Nakhjiri et al. 2001; Mao et al. 2007), multiply, survive in the epithelial cells and spread into other cells (Yilmaz et al. 2006), promote cell survival and increase cell proliferation (Kuboniwa et al. 2008). *P. gingivalis* may also induce COX-2 expression (Kuramitsu et al. 2002) and increase production of TNF-α, IL-6, IL-8, and IL-1b (Andrian et al. 2004). It was also shown that *P. gingivalis* contains proteins and lipopolysaccharides on its outer surface that can stimulate human fibroblasts to proliferate (Mihara et al. 1993; Takemura et al. 1998; Putnins et al. 2002). Based on results of these studies, Katz et al. (2011) declared a hypothesis that *P. gingivalis* may cause oral cancer and revealed that *P. gingivalis* induced neoplastic transformation of the epithelium and is present in the oral squamous cell carcinoma tissues in higher concentration than in normal tissues.

As there are only fundamental investigations devoted to the role of these infectious agents in the carcinogenesis, the conduction of the epidemiological studies are necessary, especially for the *B. henselae* and *P. gingivalis*. At the moment, it is feasible to include them in the extended list of the infectious agents that may be a cause of cancer development but are not placed in the IARC roster.

References

Abdulamir AS, Hafidh RR, Mahdi LK et al (2009) Investigation into the controversial association of Streptococcus gallolyticus with colorectal cancer and adenoma. BMC Cancer 9:403

Abrams JT, Vonderheid EC, Kolbe S et al (1999) Sézary T-cell activating factor is a *Chlamydia pneumoniae*-associated protein. Clin Diagn Lab Immunol 6(6):895–905

Abrams JT, Balin BJ, Vonderheid EC (2001) Association between Sézary T cell-activating factor, *Chlamydia pneumoniae*, and cutaneous T cell lymphoma. Ann N Y Acad Sci 941:69–85

Abramson DH, Rollins I, Coleman M (2005) Periocular mucosa-associated lymphoid/low grade lymphomas: treatment with antibiotics. Am J Ophthalmol 140(4):729–730

Ackerman Z, Eliakim R, Stalnikowicz R (1995) Spontaneous peritonitis caused by *Streptococcus bovis*: search for colonic neoplasia. J Clin Gastroenterol 21(3):263

Afek S, Sperber AD, Almog Y (2004) Carcinoma of the colon presenting as *Streptococcus salivarius* sepsis. J Clin Gastroenterol 38(1):86–87

Aigelsreiter A, Leitner E, Deutsch AJ et al (2008) Chlamydia psittaci in MALT lymphomas of ocular adnexals: the Austrian experience. Leuk Res 32(8):1292–1294

Aigelsreiter A, Gerlza T, Deutsch AJ et al (2011) Chlamydia psittaci infection in nongastrointestinal extranodal MALT lymphomas and their precursor lesions. Am J Clin Pathol 135(1):70–75

Aizawa K, Kaminishi Y, Kawamata N et al (2006) Simultaneously performed mitral valve replacement and Hartmann's operation for infectious endocarditis from *Streptococcus bovis* and rectal cancer. Jpn J Thorac Cardiovasc Surg 54(9):413–415

Alaniz Sánchez A, Flores Martínez A, León Vistrain MC et al (1995) *Chlamydia trachomatis* and cervix dysplasia. Ginecol Obstet Mex 63:377–381

Alavanja MC, Brownson RC, Boice JD Jr et al (1992) Preexisting lung disease and lung cancer among nonsmoking women. Am J Epidemiol 136(6):623–632

Alazmi W, Bustamante M, O'Loughlin C et al (2006) The association of *Streptococcus bovis* bacteremia and gastrointestinal diseases: a retrospective analysis. Dig Dis Sci 51(4):732–736

Alberico S, Facca MC, Quaranta M et al (1989) Infection of *Chlamydia trachomatis* and dysplasia lesions of the cervix uteri (CIN). Minerva Ginecol 41(7):343–347

Alexeyev O, Bergh J, Marklund I et al (2006) Association between the presence of bacterial 16S RNA in prostate specimens taken during transurethral resection of prostate and subsequent risk of prostate cancer (Sweden). Cancer Causes Control 17(9):1127–1133

Almeida J, Bau J, Baptista I et al (1992) Adenocarcinoma of the colon disclosed by endocarditis caused by *Streptococcus bovis*. Acta Med Port 5(6):335–337

Andreotti G, Liu E, Gao YT et al (2011) Medical history and the risk of biliary tract cancers in Shanghai, China: implications for a role of inflammation. Cancer Causes Control 22(9):1289–1296

Andrian E, Grenier D, Rouabhia M (2004) In vitro models of tissue penetration and destruction by *Porphyromonas gingivalis*. Infect Immun 72(8):4689–4698

Anttila TI, Lehtinen T, Leinonen M et al (1998) Serological evidence of an association between chlamydial infections and malignant lymphomas. Br J Haematol 103(1):150–156

Anttila T, Saikku P, Koskela P et al (2001) Serotypes of *Chlamydia trachomatis* and risk for development of cervical squamous cell carcinoma. JAMA 285(1):47–51

Anttila T, Koskela P, Leinonen M et al (2003) *Chlamydia pneumoniae* infection and the risk of female early-onset lung cancer. Int J Cancer 107(4):681–682

Anttila T, Tenkanen L, Lumme S et al (2005) Chlamydial antibodies and risk of prostate cancer. Cancer Epidemiol Biomarkers Prev 14(2):385–389

Apsingi S, Kulkarni A, Gould KF et al (2007) Late *Streptococcus bovis* infection of knee arthroplasty and its association with carcinoma of the colon: a case report. Knee Surg Sports Traumatol Arthrosc 15(6):761–762

Arnheim Dahlström L, Andersson K, Luostarinen T et al (2011) Prospective seroepidemiologic study of human papillomavirus and other risk factors in cervical cancer. Cancer Epidemiol Biomarkers Prev 20(12):2541–2550

Arzanauskiene R, Zabiela P, Sakalnikiene M (2003) *Streptococcus bovis* endocarditis – predictor of colonic carcinoma? Medicina (Kaunas) 39(2):174–176

Avenaud P, Le Bail B, Mayo K et al (2003) Natural history of *Helicobacter hepaticus* infection in conventional A/J mice, with special reference to liver involvement. Infect Immun 71(6):3667–3672

Baker LH, Mebust WK, Chin TD et al (1981) The relationship of herpes virus to carcinoma of the prostate. J Urol 125(3):370–374

Ballardini P, Incasa E, Gamberini S et al (2008) Not all acute lower back pain is benign – paravertebral abscess and colonic cancer. Am J Emerg Med 26(4):513.e3-4

Ballet M, Gevigney G, Garé JP et al (1995) Infective endocarditis due to *Streptococcus bovis*. A report of 53 cases. Eur Heart J 16(12):1975–1980

Bansal P, Sonnenberg A (1996) Comorbid occurrence of cholelithiasis and gastrointestinal cancer. Eur J Gastroenterol Hepatol 8(10):985–988

Barcelos AM, Teixeira MA, Alves LS et al (2010) Infectious endocarditis due to *Streptococcus bovis* in a patient with colon carcinoma. Arq Bras Cardiol 95(3):e88–e90

Bargiacchi O, Salerno AM, Brondolo R et al (2010) Whipple's disease in an AIDS patient with Kaposi's sarcoma. Report of a case. Infez Med 18(3):198

Barykova IA, Shmarov MM, Logunov DI et al (2010) Identification of Mycoplasma in patients with suspected prostate cancer. Zh Mikrobiol Epidemiol Immunobiol 4:81–85

Barykova YA, Logunov DY, Shmarov MM et al (2011) Association of *Mycoplasma hominis* infection with prostate cancer. Oncotarget 2(4):289–297

Beadsworth M, Waitt C, Jenkins N et al (2005) A retained intravascular medical device and *Streptococcus bovis* infection, providing early diagnosis of an occult malignancy. J Infect 50(2):165–167

Beeching NJ, Christmas TI, Ellis-Pegler RB et al (1985) *Streptococcus bovis* bacteraemia requires rigorous exclusion of colonic neoplasia and endocarditis. Q J Med 56(220):439–450

Belinkie SA, Narayanan NC, Russell JC et al (1983) Splenic abscess associated with *Streptococcus bovis* septicemia and neoplastic lesions of the colon. Dis Colon Rectum 26(12):823–824

Beltinger J, del Buono J, Skelly MM et al (2008) Disruption of colonic barrier function and induction of mediator release by strains of *Campylobacter jejuni* that invade epithelial cells. World J Gastroenterol 14(48):7345–7352

Biarc J, Nguyen IS, Pini A et al (2004) Carcinogenic properties of proteins with pro-inflammatory activity from *Streptococcus infantarius* (formerly *S. bovis*). Carcinogenesis 25(8):1477–1484

Bjørge T, Hakulinen T, Engeland A et al (1997) A prospective, seroepidemiological study of the role of human papillomavirus in esophageal cancer in Norway. Cancer Res 57(18):3989–3992

Bleibel W, D'Silva K, Elhorr A (2007) *Streptococcus bovis* endophthalmitis: a unique presentation of colon cancer. Dig Dis Sci 52(9):2336–2339

Böggemeyer E, Stehle T, Schaible UE et al (1994) *Borrelia burgdorferi* upregulates the adhesion molecules E-selectin, P-selectin, ICAM-1 and VCAM-1 on mouse endothelioma cells in vitro. Cell Adhes Commun 2(2):145–157

Bogle MA, Riddle CC, Triana EM et al (2005) Primary cutaneous B-cell lymphoma. J Am Acad Dermatol 53(3):479–484

Boleij A, Roelofs R, Schaeps RM et al (2010) Increased exposure to bacterial antigen RpL7/L12 in early stage colorectal cancer patients. Cancer 116(17):4014–4022

Boleij A, Muytjens CM, Bukhari SI et al (2011a) Novel clues on the specific association of *Streptococcus gallolyticus* subsp. *gallolyticus* with colorectal cancer. J Infect Dis 203(8):1101–1109

Boleij A, van Gelder MM, Swinkels DW et al (2011b) Clinical importance of *Streptococcus gallolyticus* infection among colorectal cancer patients: systematic review and meta-analysis. Clin Infect Dis 53(9):870–878

Boleij A, Roelofs R, Danne C et al (2012) Selective antibody response to *Streptococcus gallolyticus* pilus proteins in colorectal cancer patients. Cancer Prev Res (Phila) 5(2):260–265

Borisov I, Maĭnkhard K (1995) The relationship between genital chlamydial infection and the presence of cervical intraepithelial neoplasia. Akush Ginekol (Sofiia) 34(3):39–40

Bosch FX, Castellsagué X, Muñoz N et al (1996) Male sexual behavior and human papillomavirus DNA: key risk factors for cervical cancer in Spain. J Natl Cancer Inst 88(15):1060–1067

Boyle DC, Barton SE, Uthayakumar S et al (2003) Is bacterial vaginosis associated with cervical intraepithelial neoplasia? Int J Gynecol Cancer 13(2):159–163

Brauner A, Brandt L, Frisan T et al (2010) Is there a risk of cancer development after Campylobacter infection? Scand J Gastroenterol 45(7–8):893–897

Brenner AV, Wang Z, Kleinerman RA et al (2001) Previous pulmonary diseases and risk of lung cancer in Gansu Province, China. Int J Epidemiol 30(1):118–124

Brenner DR, Hung RJ, Tsao MS et al (2010) Lung cancer risk in never-smokers: a population-based case-control study of epidemiologic risk factors. BMC Cancer 10:285

Brenner DR, McLaughlin JR, Hung RJ (2011) Previous lung diseases and lung cancer risk: a systematic review and meta-analysis. PLoS One 6(3):e17479

Brownson RC, Alavanja MC (2000) Previous lung disease and lung cancer risk among women (United States). Cancer Causes Control 11(9):853–858

Buckley DA, Murphy A, Dervan P et al (1998) Persistent infection of the chin with an unusual skin pathogen (*Streptococcus milleri*): a sign of intraoral carcinoma. Clin Exp Dermatol 23(1):35–37

Bułhak-Kozioł V, Zdrodowska-Stefanow B, Ostaszewska-Puchalska I et al (2007) Prevalence of *Chlamydia trachomatis* infection in women with cervical lesions. Adv Med Sci 52:179–181

Burns CA, McCaughey R, Lauter CB (1985) The association of *Streptococcus bovis* fecal carriage and colon neoplasia: possible relationship with polyps and their premalignant potential. Am J Gastroenterol 80(1):42–46

Cai JP, Cheng JW, Ma XY et al (2012) Lack of association of conjunctival MALT lymphoma with Chlamydiae or *Helicobacter pylori* in a cohort of Chinese patients. Med Sci Monit 18(2):BR84–BR88

Cailliez-Tomasi JP, Avisse C, Baehrel B et al (1997) Combination of colorectal tumor and *Streptococcus bovis* endocarditis. Apropos of 3 cases. J Chir (Paris) 134(2):69–72

Cardillo MR (1985) Association of human papilloma virus and *Chlamydia trachomatis* infections with incidence cervical neoplasia. Eur J Gynaecol Oncol 6(3):218–221

Carratelli CR, Rizzo A, Catania MR et al (2002) *Chlamydia pneumoniae* infections prevent the programmed cell death on THP-1 cell line. FEMS Microbiol Lett 215(1):69–74

Carta G, di Stefano L, Scimia G et al (1994) *Chlamydia trachomatis* infection and its correlation with cervical preoncogenesis. Minerva Ginecol 46(12):651–655

Cassidy A, Myles JP, van Tongeren M et al (2008) The LLP risk model: an individual risk prediction model for lung cancer. Br J Cancer 98(2):270–276

Castle PE, Escoffery C, Schachter J et al (2003) *Chlamydia trachomatis*, herpes simplex virus 2, and human T-cell lymphotrophic virus type 1 are not associated with grade of cervical neoplasia in Jamaican colposcopy patients. Sex Transm Dis 30(7):575–580

Caygill CP, Braddick M, Hill MJ et al (1995) The association between typhoid carriage, typhoid infection and subsequent cancer at a number of sites. Eur J Cancer Prev 4(2):187–193

Cerroni L, Zöchling N, Pütz B (1997) Infection by *Borrelia burgdorferi* and cutaneous B-cell lymphoma. J Cutan Pathol 24(8):457–461

Cerroni L, Höfler G, Bäck B et al (2002) Specific cutaneous infiltrates of B-cell chronic lymphocytic leukemia (B-CLL) at sites typical for *Borrelia burgdorferi* infection. J Cutan Pathol 29(3):142–147

Cevenini R, Costa S, Rumpianesi F et al (1981) Cytological and histopathological abnormalities of the cervix in genital *Chlamydia trachomatis* infections. Br J Vener Dis 57(5):334–337

Chan CC, Shen D, Mochizuki M et al (2006) Detection of *Helicobacter pylori* and *Chlamydia pneumoniae* genes in primary orbital lymphoma. Trans Am Ophthalmol Soc 104:62–70

Chang CM, Landgren O, Koshiol J et al (2012) Borrelia and subsequent risk of solid tumors and hematologic malignancies in Sweden. Int J Cancer 131(9):2208–2209

Chanudet E, Zhou Y, Bacon CM et al (2006) Chlamydia psittaci is variably associated with ocular adnexal MALT lymphoma in different geographical regions. J Pathol 209(3):344–351

Chanudet E, Adam P, Nicholson AG et al (2007) Chlamydiae and Mycoplasma infections in pulmonary MALT lymphoma. Br J Cancer 97(7):949–951

Chan-Yeung M, Koo LC, Ho JC et al (2003) Risk factors associated with lung cancer in Hong Kong. Lung Cancer 40(2):131–140

Checkoway H, DiFerdinando G, Hulka BS et al (1987) Medical, life-style, and occupational risk factors for prostate cancer. Prostate 10(1):79–88

Chiong C, Cox MR, Eslick GD (2012) Gallstone disease is associated with rectal cancer: a meta-analysis. Scand J Gastroenterol 47(5):553–564

Chmelar D, Vrtný J (2010) Endotoxin production in members of the *Bacteroides fragilis* group in relation to colorectal cancer humans. Klin Mikrobiol Infekc Lek 16(3):97–102

Chomarat P, Sipowicz MA, Diwan BA et al (1997) Distinct time courses of increase in cytochromes P450 1A2, 2A5 and glutathione S-transferases during the progressive hepatitis associated with *Helicobacter hepaticus*. Carcinogenesis 18(11):2179–2190

Chow J, Mazmanian SK (2010) A pathobiont of the microbiota balances host colonization and intestinal inflammation. Cell Host Microbe 7(4):265–276

Chung KT, Fulk GE, Slein MW (1975) Tryptophanase of fecal flora as a possible factor in the etiology of colon cancer. J Natl Cancer Inst 54(5):1073–1078

Cocco P, Rice CH, Chen JQ et al (2000) Non-malignant respiratory diseases and lung cancer among Chinese workers exposed to silica. J Occup Environ Med 42(6):639–644

Cohen RJ, Shannon BA, McNeal JE et al (2005) Propionibacterium acnes associated with inflammation in radical prostatectomy specimens: a possible link to cancer evolution? J Urol 173(6):1969–1974

Conde Martel A, Martín Pérez P, Suárez Ortega S et al (2002) Hepatic abscess caused by *Streptococcus intermedius* as first manifestation of renal carcinoma. Rev Esp Enferm Dig 94(1):45–46

Copeland LG, Malster MG (1993) Suspicious mind: the association between *Streptococcus bovis* endocarditis and carcinoma of the colon. Postgrad Med J 69(809):241

Corredoira JC, Alonso MP, García JF et al (2005) Clinical characteristics and significance of *Streptococcus salivarius* bacteremia and *Streptococcus bovis* bacteremia: a prospective 16-year study. Eur J Clin Microbiol Infect Dis 24(4):250–255

Corredoira J, Alonso MP, Coira A et al (2008a) Association between *Streptococcus infantarius* (formerly *S. bovis* II/1) bacteremia and noncolonic cancer. J Clin Microbiol 46(4):1570

Corredoira J, Alonso MP, Coira A et al (2008b) Characteristics of *Streptococcus bovis* endocarditis and its differences with *Streptococcus viridans* endocarditis. Eur J Clin Microbiol Infect Dis 27(4):285–291

Costa CP, Kirschning CJ, Busch D et al (2002) Role of chlamydial heat shock protein 60 in the stimulation of innate immune cells by *Chlamydia pneumoniae*. Eur J Immunol 32(9):2460–2470

Daibata M, Nemoto Y, Togitani K et al (2006) Absence of *Chlamydia psittaci* in ocular adnexal lymphoma from Japanese patients. Br J Haematol 132(5):651–652

Daling JR, Weiss NS, Hislop TG et al (1987) Sexual practices, sexually transmitted diseases, and the incidence of anal cancer. N Engl J Med 317(16):973–977

de Cremoux P, Subtil A, Ferreri AJ et al (2006) Re: evidence for an association between *Chlamydia psittaci* and ocular adnexal lymphomas. J Natl Cancer Inst 98(5):365–366

de Sanjosé S, Muñoz N, Bosch FX et al (1994) Sexually transmitted agents and cervical neoplasia in Colombia and Spain. Int J Cancer 56(3):358–363

Dehio C (2003) Recent progress in understanding Bartonella-induced vascular proliferation. Curr Opin Microbiol 6(1):61–65

Dennis LK, Lynch CF, Torner JC (2002) Epidemiologic association between prostatitis and prostate cancer. Urology 60(1):78–83

Dennis LK, Coughlin JA, McKinnon BC et al (2009) Sexually transmitted infections and prostate cancer among men in the U.S. military. Cancer Epidemiol Biomarkers Prev 18(10): 2665–2671

Devis A, Dony A, De Boelpaepe F et al (1989) *Streptococcus bovis* septicemia and colonic cancer. Acta Chir Belg 89(1):58–60

Dicksved J, Lindberg M, Rosenquist M et al (2009) Molecular characterization of the stomach microbiota in patients with gastric cancer and in controls. J Med Microbiol 58(Pt 4):509–516

Dieleman LA, Arends A, Tonkonogy SL et al (2000) *Helicobacter hepaticus* does not induce or potentiate colitis in interleukin-10-deficient mice. Infect Immun 68(9):5107–5113

Dillner J, Lenner P, Lehtinen M et al (1994) A population-based seroepidemiological study of cervical cancer. Cancer Res 54(1):134–141

Dillner J, Lehtinen M, Björge T et al (1997) Prospective seroepidemiologic study of human papillomavirus infection as a risk factor for invasive cervical cancer. J Natl Cancer Inst 89(17):1293–1299

Diwan BA, Ward JM, Ramljak D et al (1997) Promotion by *Helicobacter hepaticus*-induced hepatitis of hepatic tumors initiated by N-nitrosodimethylamine in male A/JCr mice. Toxicol Pathol 25(6):597–605

Diwan BA, Ramakrishna G, Anderson LM et al (2000) Overexpression of Grb2 in inflammatory lesions and preneoplastic foci and tumors induced by N-nitrosodimethylamine in *Helicobacter hepaticus*-infected and -noninfected A/J mice. Toxicol Pathol 28(4):548–554

Diwan BA, Sipowicz M, Logsdon D et al (2008) Marked liver tumorigenesis by *Helicobacter hepaticus* requires perinatal exposure. Environ Health Perspect 116(10):1352–1356

Dowd DE, Loftus EF (1985) A case of *Streptococcus bovis* endocarditis and colonic carcinoma. Ir Med J 78(2):52–53

Duncan ME, Pelzer A, Tibaux G et al (1993) Is cervical cancer in Ethiopian women the result of a sexually transmitted disease? Int Surg 78(2):134–140

Dunham WR, Simpson JH, Feest TG et al (1980) *Streptococcus bovis* endocarditis and colorectal disease. Lancet 1(8165):421–422

Dutta U, Garg PK, Kumar R et al (2000) Typhoid carriers among patients with gallstones are at increased risk for carcinoma of the gallbladder. Am J Gastroenterol 95(3):784–787

Duval X, Papastamopoulos V, Longuet P et al (2001) Definite *Streptococcus bovis* endocarditis: characteristics in 20 patients. Clin Microbiol Infect 7(1):3–10

Ebnet K, Simon MM, Shaw S (1996) Regulation of chemokine gene expression in human endothelial cells by proinflammatory cytokines and *Borrelia burgdorferi*. Ann N Y Acad Sci 797:107–117

Ebnet K, Brown KD, Siebenlist UK et al (1997) *Borrelia burgdorferi* activates nuclear factor-kappa B and is a potent inducer of chemokine and adhesion molecule gene expression in endothelial cells and fibroblasts. J Immunol 158(7):3285–3292

Ekiel AM, Friedek DA, Romanik MK et al (2009) Occurrence of Ureaplasma parvum and Ureaplasma urealyticum in women with cervical dysplasia in Katowice, Poland. J Korean Med Sci 24(6):1177–1181

Ellmerich S, Schöller M, Duranton B et al (2000) Promotion of intestinal carcinogenesis by *Streptococcus bovis*. Carcinogenesis 21(4):753–756

el-Zayadi A, Ghoneim M, Kabil SM et al (1991) Bile duct carcinoma in Egypt: possible etiological factors. Hepatogastroenterology 38(4):337–340

Engels EA, Shen M, Chapman RS et al (2009) Tuberculosis and subsequent risk of lung cancer in Xuanwei, China. Int J Cancer 124(5):1183–1187

Erdman SE, Poutahidis T, Tomczak M et al (2003a) CD4+ CD25+ regulatory T lymphocytes inhibit microbially induced colon cancer in Rag2-deficient mice. Am J Pathol 162(2):691–702

Erdman SE, Rao VP, Poutahidis T et al (2003b) CD4(+)CD25(+) regulatory lymphocytes require interleukin 10 to interrupt colon carcinogenesis in mice. Cancer Res 63(18):6042–6050

Erdman SE, Rao VP, Poutahidis T et al (2009) Nitric oxide and TNF-alpha trigger colonic inflammation and carcinogenesis in *Helicobacter hepaticus*-infected, Rag2-deficient mice. Proc Natl Acad Sci USA 106(4):1027–1032

Fassi Fehri L, Mak TN, Laube B et al (2011) Prevalence of *Propionibacterium acnes* in diseased prostates and its inflammatory and transforming activity on prostate epithelial cells. Int J Med Microbiol 301(1):69–78

Fernández-Ruiz M, Villar-Silva J, Llenas-García J et al (2010) *Streptococcus bovis* bacteraemia revisited: clinical and microbiological correlates in a contemporary series of 59 patients. J Infect 61(4):307–313

Ferrari A, Botrugno I, Bombelli E (2008) Colonoscopy is mandatory after *Streptococcus bovis* endocarditis: a lesson still not learned. Case report. World J Surg Oncol 6:49

Ferrera A, Baay MF, Herbrink P et al (1997) A sero-epidemiological study of the relationship between sexually transmitted agents and cervical cancer in Honduras. Int J Cancer 73(6):781–785

Ferreri AJ, Guidoboni M, Ponzoni M et al (2004) Evidence for an association between *Chlamydia psittaci* and ocular adnexal lymphomas. J Natl Cancer Inst 96(8):586–594

Ferreri AJ, Ponzoni M, Guidoboni M et al (2005) Regression of ocular adnexal lymphoma after *Chlamydia psittaci*-eradicating antibiotic therapy. J Clin Oncol 23(22):5067–5073

Ferreri AJ, Dognini GP, Ponzoni M et al (2008a) *Chlamydia-psittaci*-eradicating antibiotic therapy in patients with advanced-stage ocular adnexal MALT lymphoma. Ann Oncol 19(1):194–195

Ferreri AJ, Dolcetti R, Dognini GP et al (2008b) *Chlamydophila psittaci* is viable and infectious in the conjunctiva and peripheral blood of patients with ocular adnexal lymphoma: results of a single-center prospective case-control study. Int J Cancer 123(5):1089–1093

Fischer N (2002) *Chlamydia trachomatis* infection in cervical intraepithelial neoplasia and invasive carcinoma. Eur J Gynaecol Oncol 23(3):247–250

Florescu M, Benea DC, Cerin G et al (2009) Aorto-pulmonary fistula as a late complication of multiple valve replacement after *Streptococcus bovis* endocarditis in a patient with occult colon carcinoma. J Clin Ultrasound 37(6):369–373

Foltz CJ, Fox JG, Cahill R et al (1998) Spontaneous inflammatory bowel disease in multiple mutant mouse lines: association with colonization by *Helicobacter hepaticus*. Helicobacter 3(2):69–78

Fox JG, Dewhirst FE, Tully JG et al (1994) *Helicobacter hepaticus* sp. nov., a microaerophilic bacterium isolated from livers and intestinal mucosal scrapings from mice. J Clin Microbiol 32(5):1238–1245

Fox JG, Yan LL, Dewhirst FE et al (1995) *Helicobacter bilis* sp. nov., a novel *Helicobacter* species isolated from bile, livers, and intestines of aged, inbred mice. J Clin Microbiol 33(2):445–454

Fox JG, Li X, Yan L et al (1996a) Chronic proliferative hepatitis in A/JCr mice associated with persistent *Helicobacter hepaticus* infection: a model of helicobacter-induced carcinogenesis. Infect Immun 64(5):1548–1558

Fox JG, Yan L, Shames B et al (1996b) Persistent hepatitis and enterocolitis in germfree mice infected with *Helicobacter hepaticus*. Infect Immun 64(9):3673–3681

Fox JG, Dewhirst FE, Shen Z et al (1998) Hepatic *Helicobacter* species identified in bile and gallbladder tissue from Chileans with chronic cholecystitis. Gastroenterology 114(4):755–763

Fox JG, Feng Y, Theve EJ et al (2010) Gut microbes define liver cancer risk in mice exposed to chemical and viral transgenic hepatocarcinogens. Gut 59(1):88–97

Friedman E, Elian D, Eisenstein Z (1986) *Streptococcus bovis* bacteremia and underlying gastrointestinal neoplasms. Med Pediatr Oncol 14(6):313–315

Friedrich IA, Wormser GP, Gottfried EB (1982) The association of remote *Streptococcus bovis* bacteremia with colonic neoplasia. Am J Gastroenterol 77(2):82–84

Frisch M, Jørgensen BB, Friis S et al (1996) Syphilis and the risk of penis cancer. Sex Transm Dis 23(6):471–474

Fühler M, Ottmann KW, Tronnier M (2010) Cutaneous marginal zone lymphoma (SALT) and infection with *Borrelia burgdorferi*. Hautarzt 61(2):145–147

Furgyik S (1985) Increased frequency of carcinoma of the uterine cervix in women married to men with a previous gonorrheal infection. An epidemiological study. Acta Obstet Gynecol Scand 64(3):245–249

Furgyik S (1987) Gonorrhea infection as a risk indicator for cervix cancer. Geburtshilfe Frauenheilkd 47(5):320–323

Furgyik S, Astedt B (1980) Gonorrheal infection followed by an increased frequency of cervical carcinoma. Acta Obstet Gynecol Scand 59(6):521–524

Galeone C, Pelucchi C, La Vecchia C et al (2008) Indoor air pollution from solid fuel use, chronic lung diseases and lung cancer in Harbin, Northeast China. Eur J Cancer Prev 17(5):473–478

Galobardes B, Davey Smith G, Jeffreys M et al (2005) Acne in adolescence and cause-specific mortality: lower coronary heart disease but higher prostate cancer mortality: the Glasgow Alumni Cohort Study. Am J Epidemiol 161(12):1094–1101

Gao YT, Blot WJ, Zheng W et al (1987) Lung cancer among Chinese women. Int J Cancer 40(5):604–609

Garbe C, Stein H, Gollnick H et al (1988) Cutaneous B cell lymphoma in chronic *Borrelia burgdorferi* infection. Report of 2 cases and a review of the literature. Hautarzt 39(11):717–726

Garbe C, Stein H, Dienemann D et al (1991) *Borrelia burgdorferi*-associated cutaneous B cell lymphoma: clinical and immunohistologic characterization of four cases. J Am Acad Dermatol 24(4):584–590

García A, Ihrig MM, Fry RC et al (2008) Genetic susceptibility to chronic hepatitis is inherited codominantly in *Helicobacter hepaticus*-infected AB6F1 and B6AF1 hybrid male mice, and progression to hepatocellular carcinoma is linked to hepatic expression of lipogenic genes and immune function-associated networks. Infect Immun 76(5):1866–1876

García A, Zeng Y, Muthupalani S et al (2011) *Helicobacter hepaticus* – induced liver tumor promotion is associated with increased serum bile acid and a persistent microbial-induced immune response. Cancer Res 71(7):2529–2540

Garcia-Porrua C, Gonzalez-Gay MA, Monterroso JR et al (2000) Septic arthritis due to *Streptococcus bovis* as presenting sign of 'silent' colon carcinoma. Rheumatology (Oxford) 39(3):338–339

Ge Z, Rogers AB, Feng Y et al (2007) Bacterial cytolethal distending toxin promotes the development of dysplasia in a model of microbially induced hepatocarcinogenesis. Cell Microbiol 9(8):2070–2080

Genta PR, Carneiro L, Genta EN (1998) *Streptococcus bovis* bacteremia: unusual complications. South Med J 91(12):1167–1168

Ger LP, Hsu WL, Chen KT et al (1993) Risk factors of lung cancer by histological category in Taiwan. Anticancer Res 13(5A):1491–1500

Giannitsioti E, Chirouze C, Bouvet A et al (2007) Characteristics and regional variations of group D streptococcal endocarditis in France. Clin Microbiol Infect 13(8):770–776

Gil Lasa I, Montalvo Ollobarren II, Mensa Landaburu G et al (1995) Liver abscess caused by *Streptococcus anginosus*. Rev Esp Enferm Dig 87(11):821–823

Giles GG, Severi G, English DR et al (2003) Early growth, adult body size and prostate cancer risk. Int J Cancer 103(2):241–245

Gillen CD, Coddington R, Monteith PG (1993) Extraintestinal lymphoma in association with Whipple's disease. Gut 34(11):1627–1629

Giuliano AR, Papenfuss M, Abrahamsen M et al (2001) Human papillomavirus infection at the United States-Mexico border: implications for cervical cancer prevention and control. Cancer Epidemiol Biomarkers Prev 10(11):1129–1136

Gleeson DC, Fielding J, Heath DA (1983) *Streptococcus milleri* liver abscesses associated with leiomyosarcoma of the ileum. Postgrad Med J 59(691):323–324

Gold JS, Bayar S, Salem RR (2004) Association of *Streptococcus bovis* bacteremia with colonic neoplasia and extracolonic malignancy. Arch Surg 139(7):760–765

González Sánchez JL, Flores Avilés Y, Gómez Campos G et al (1995) Association of *Chlamydia trachomatis* and human papilloma virus as predisposing factors in cervical intraepithelial neoplasia. Ginecol Obstet Mex 63:422–426

González-Juanatey C, González-Gay MA, Llorca J et al (2003) Infective endocarditis due to *Streptococcus bovis* in a series of nonaddict patients: clinical and morphological characteristics of 20 cases and review of the literature. Can J Cardiol 19(10):1139–1145

Gonzalez-Quintela A, Martínez-Rey C, Castroagudín JF et al (2001) Prevalence of liver disease in patients with *Streptococcus bovis* bacteraemia. J Infect 42(2):116–119

Goodlad JR, Davidson MM, Hollowood K et al (2000a) *Borrelia burgdorferi*-associated cutaneous marginal zone lymphoma: a clinicopathological study of two cases illustrating the temporal progression of B. burgdorferi-associated B-cell proliferation in the skin. Histopathology 37(6):501–508

Goodlad JR, Davidson MM, Hollowood K et al (2000b) Primary cutaneous B-cell lymphoma and *Borrelia burgdorferi* infection in patients from the Highlands of Scotland. Am J Surg Pathol 24(9):1279–1285

Goodwin AC, Destefano Shields CE, Wu S et al (2011) Polyamine catabolism contributes to enterotoxigenic *Bacteroides fragilis*-induced colontumorigenesis. Proc Natl Acad Sci USA 108(37):15354–15359

Gorlova OY, Zhang Y, Schabath MB et al (2006) Never smokers and lung cancer risk: a case-control study of epidemiological factors. Int J Cancer 118(7):1798–1804

Goteri G, Ranaldi R, Simonetti O et al (2007) Clinicopathological features of primary cutaneous B-cell lymphomas from an academic regional hospital in central Italy: no evidence of *Borrelia burgdorferi* association. Leuk Lymphoma 48(11):2184–2188

Goumas PD, Naxakis SS, Rentzis GA et al (1997) Lateral neck abscess caused by *Streptococcus bovis* in a patient with undiagnosed colon cancer. J Laryngol Otol 111(7):666–668

Gracia E, Froesch P, Mazzucchelli L et al (2007) Low prevalence of *Chlamydia psittaci* in ocular adnexal lymphomas from Cuban patients. Leuk Lymphoma 48(1):104–108

Grünberger B, Hauff W, Lukas J et al (2006) 'Blind' antibiotic treatment targeting Chlamydia is not effective in patients with MALT lymphoma of the ocular adnexa. Ann Oncol 17(3):484–487

Gruner U, Goesch P, Donner A et al (2001) Whipple disease and non-Hodgkin lymphoma. Z Gastroenterol 39(4):305–309

Guaschino S, Stola E, Spinillo A (1988) *Chlamydia trachomatis* and cervical intraepithelial neoplasia. Clin Exp Obstet Gynecol 15(3):98–101

Guijon F, Paraskevas M, Rand F et al (1992) Vaginal microbial flora as a cofactor in the pathogenesis of uterine cervical intraepithelial neoplasia. Int J Gynaecol Obstet 37(3):185–191

Gupta A, Madani R, Mukhtar H (2010) *Streptococcus bovis* endocarditis, a silent sign for colonic tumour. Colorectal Dis 12(3):164–171

Hailey JR, Haseman JK, Bucher JR et al (1998) Impact of *Helicobacter hepaticus* infection in B6C3F1 mice from twelve National Toxicology Program two-year carcinogenesis studies. Toxicol Pathol 26(5):602–611

Hakama M, Lehtinen M, Knekt P et al (1993) Serum antibodies and subsequent cervical neoplasms: a prospective study with 12 years of follow-up. Am J Epidemiol 137(2):166–170

Hakama M, Luostarinen T, Hallmans G et al (2000) Joint effect of HPV16 with *Chlamydia trachomatis* and smoking on risk of cervical cancer: antagonism or misclassification (Nordic countries). Cancer Causes Control 11(9):783–790

Hamada T, Yokota K, Ayada K et al (2009) Detection of *Helicobacter hepaticus* in human bile samples of patients with biliary disease. Helicobacter 14(6):545–551

Hare MJ, Taylor-Robinson D, Cooper P (1982) Evidence for an association between *Chlamydia trachomatis* and cervical intraepithelial neoplasia. Br J Obstet Gynaecol 89(6):489–492

Harnik IG (2007) Pyogenic liver abscess presenting after malignant polypectomy. Dig Dis Sci 52(12):3524–3525

Hatanaka K, Miyagishima T, Kamata T et al (2000) Occurrence of angioimmunoblastic T-cell lymphoma six months after onset of Lyme disease. Rinsho Ketsueki 41(12):1273–1276

Hayes RB, Pottern LM, Strickler H et al (2000) Sexual behaviour, STDs and risks for prostate cancer. Br J Cancer 82(3):718–725

Heggie AD, Wentz WB, Reagan JW et al (1986) Roles of cytomegalovirus and *Chlamydia trachomatis* in the induction of cervical neoplasia in the mouse. Cancer Res 46(10):5211–5214

Hernandez-Vaquero D, Cuervo-Olay MC, Suarez-Vazquez A (2008) Association of *Streptococcus bovis* haematolgenous infection in total knee replacement with large intestine carcinoma. Orthopedics 31(1):92

Herrero R, Brinton LA, Reeves WC et al (1990) Sexual behavior, venereal diseases, hygiene practices, and invasive cervical cancer in a high-risk population. Cancer 65(2):380–386

Herrero IA, Rouse MS, Piper KE et al (2002) Reevaluation of *Streptococcus bovis* endocarditis cases from 1975 to 1985 by 16S ribosomal DNA sequence analysis. J Clin Microbiol 40(10):3848–3850

Heshmat MY, Kovi J, Herson J et al (1975) Epidemiologic association between Gonorrhea and prostatic carcinoma. Urology 6(4):457–460

Heuvers ME, Aerts JG, Hegmans JP et al (2012) History of tuberculosis as an independent prognostic factor for lung cancer survival. Lung Cancer 76(3):452–456

Hiatt RA, Armstrong MA, Klatsky AL et al (1994) Alcohol consumption, smoking, and other risk factors and prostate cancer in a large health plan cohort in California (United States). Cancer Causes Control 5(1):66–72

Hinds MW, Cohen HI, Kolonel LN (1982) Tuberculosis and lung cancer risk in nonsmoking women. Am Rev Respir Dis 125(6):776–778

Hinkula M, Pukkala E, Kyyrönen P et al (2004) A population-based study on the risk of cervical cancer and cervical intraepithelial neoplasia among grand multiparous women in Finland. Br J Cancer 90(5):1025–1029

Hirose K, Isogai E, Mizugai H et al (1996) Adhesion of *Porphyromonas gingivalis* fimbriae to human gingival cell line Ca9-22. Oral Microbiol Immunol 11(6):402–406

Hoen B, Briançon S, Delahaye F et al (1994) Tumors of the colon increase the risk of developing *Streptococcus bovis* endocarditis: case-control study. Clin Infect Dis 19(2):361–362

Honda GD, Bernstein L, Ross RK et al (1988) Vasectomy, cigarette smoking, and age at first sexual intercourse as risk factors for prostate cancer in middle-aged men. Br J Cancer 57(3): 326–331

Hossenbux K, Dale BA, Walls AD (1983) *Streptococcus bovis* endocarditis and colonic carcinoma: a neglected association. Br Med J (Clin Res Ed) 287(6384):21

Hrbacek J, Urban M, Hamsikova E et al (2011) Serum antibodies against genitourinary infectious agents in prostate cancer and benign prostate hyperplasia patients: a case-control study. BMC Cancer 11:53

Hsieh CC, Thanos A, Mitropoulos D et al (1999a) Risk factors for prostate cancer: a case-control study in Greece. Int J Cancer 80(5):699–703

Hsieh CY, You SL, Kao CL et al (1999b) Reproductive and infectious risk factors for invasive cervical cancer in Taiwan. Anticancer Res 19(5C):4495–4500

Hsing AW, Gao YT, Han TQ et al (2007) Gallstones and the risk of biliary tract cancer: a population-based study in China. Br J Cancer 97(11):1577–1582

Hu W, Gu M (1995) A study on the relationship between cervical carcinoma, dysplasia and sexually transmitted diseases. Zhonghua Fu Chan Ke Za Zhi 30(9):547–549

Huang WY, Hayes R, Pfeiffer R et al (2008) Sexually transmissible infections and prostate cancer risk. Cancer Epidemiol Biomarkers Prev 17(9):2374–2381

Husain A, Roberts D, Pro B et al (2007) Meta-analyses of the association between *Chlamydia psittaci* and ocular adnexal lymphoma and the response of ocular adnexal lymphoma to antibiotics. Cancer 110(4):809–815

Idahl A, Lundin E, Elgh F et al (2010) *Chlamydia trachomatis*, *Mycoplasma genitalium*, *Neisseria gonorrhoeae*, human papillomavirus, and polyomavirus are not detectable in human tissue with epithelial ovarian cancer, borderline tumor, or benign conditions. Am J Obstet Gynecol 202(1):71.e1-6

Idahl A, Lundin E, Jurstrand M et al (2011) *Chlamydia trachomatis* and *Mycoplasma genitalium* plasma antibodies in relation to epithelial ovarian tumors. Infect Dis Obstet Gynecol 2011:824627

Ilić M, Vlajinac H, Marinković J (1996) Case-control study of risk factors for prostate cancer. Br J Cancer 74(10):1682–1686

Jackson LA, Wang SP, Nazar-Stewart V et al (2000) Association of *Chlamydia pneumoniae* immunoglobulin A seropositivity and risk of lung cancer. Cancer Epidemiol Biomarkers Prev 9(11):1263–1266

Jean SS, Teng LJ, Hsueh PR et al (2004) Bacteremic *Streptococcus bovis* infections at a university hospital, 1992–2001. J Formos Med Assoc 103(2):118–123

Jelić S, Filipović-Ljesković I (1999) Positive serology for Lyme disease borrelias in primary cutaneous B-cell lymphoma: a study in 22 patients; is it a fortuitous finding? Hematol Oncol 17(3):107–116

Jha PK, Beral V, Peto J et al (1993) Antibodies to human papillomavirus and to other genital infectious agents and invasive cervical cancer risk. Lancet 341(8853):1116–1118

Johansen C, Mellemkjaer L, Frisch M et al (2001) Risk for anogenital cancer and other cancer among women hospitalized with gonorrhea. Acta Obstet Gynecol Scand 80(8):757–761

Kalimo K, Terho P, Honkonen E et al (1981) *Chlamydia trachomatis* and herpes simplex virus IgA antibodies in cervical secretions of patients with cervical atypia. Br J Obstet Gynaecol 88(11):1130–1134

Kampe CE, Vovan T, Alim A et al (1995) *Streptococcus sanguis* bacteremia and colorectal cancer: a case report. Med Pediatr Oncol 24(1):67–68

Kash N, Fink-Puches R, Cerroni L (2011) Cutaneous manifestations of B-cell chronic lymphocytic leukemia associated with *Borrelia burgdorferi* infection showing a marginal zone B-cell lymphoma-like infiltrate. Am J Dermatopathol 33(7):712–715

Katz J, Onate MD, Pauley KM et al (2011) Presence of *Porphyromonas gingivalis* in gingival squamous cell carcinoma. Int J Oral Sci 3(4):209–215

Kempf VA, Volkmann B, Schaller M et al (2001) Evidence of a leading role for VEGF in *Bartonella henselae*-induced endothelial cell proliferations. Cell Microbiol 3(9):623–632

Kewal N, Seneviratne BI, Wilkinson RK et al (1983) Asymptomatic colonic carcinoma revealed by investigation of *Streptococcus bovis* bacteremia. Aust N Z J Med 13(2):173–174

Kim JM, Cho SJ, Oh YK et al (2002) Nuclear factor-kappa B activation pathway in intestinal epithelial cells is a major regulator of chemokine gene expression and neutrophil migration induced by *Bacteroides fragilis* enterotoxin. Clin Exp Immunol 130(1):59–66

Kim SY, Joo SI, Yi J et al (2010a) A case of *Streptococcus gallolyticus* subsp. gallolyticus infective endocarditis with colon cancer: identification by 16S ribosomal DNA sequencing. Korean J Lab Med 30(2):160–165

Kim YC, Ko Y, Hong SD et al (2010b) Presence of *Porphyromonas gingivalis* and plasma cell dominance in gingival tissues with periodontitis. Oral Dis 16(4):375–381

Kirby JE (2004) In vitro model of *Bartonella henselae*-induced angiogenesis. Infect Immun 72(12):7315–7317

Kirby JE, Nekorchuk DM (2002) Bartonella-associated endothelial proliferation depends on inhibition of apoptosis. Proc Natl Acad Sci USA 99(7):4656–4661

Kjaer SK, Dahl C, Engholm G et al (1992) Case-control study of risk factors for cervical neoplasia in Denmark. II. Role of sexual activity, reproductive factors, and venereal infections. Cancer Causes Control 3(4):339–348

Klein RS, Recco RA, Catalano MT et al (1977) Association of *Streptococcus bovis* with carcinoma of the colon. N Engl J Med 297(15):800–802

Klein RS, Catalano MT, Edberg SC et al (1979) *Streptococcus bovis* septicemia and carcinoma of the colon. Ann Intern Med 91(4):560–562

Ko YC, Lee CH, Chen MJ et al (1997) Risk factors for primary lung cancer among non-smoking women in Taiwan. Int J Epidemiol 26(1):24–31

Kocazeybek B (2003) Chronic *Chlamydophila pneumoniae* infection in lung cancer, a risk factor: a case-control study. J Med Microbiol 52(Pt 8):721–726

Kodama K, Massone C, Chott A et al (2005) Primary cutaneous large B-cell lymphomas: clinicopathologic features, classification, and prognostic factors in a large series of patients. Blood 106(7):2491–2497

Koh WP, Chow VT, Phoon MC et al (2005) Lack of association between chronic *Chlamydophila pneumoniae* infection and lung cancer among nonsmoking Chinese women in Singapore. Int J Cancer 114(3):502–504

Köhler U, Wuttke P (1994) Results of a case-control study of the current effect of various factors on risk of cervix cancer. 1. Factors in reproduction, sex behavior and infectious genital diseases. Zentralbl Gynakol 116(6):318–324

Kok H, Jureen R, Soon CY et al (2007) Colon cancer presenting as *Streptococcus gallolyticus* infective endocarditis. Singapore Med J 48(2):e43–e45

Koshiol J, Rotunno M, Consonni D et al (2010) Lower risk of lung cancer after multiple pneumonia diagnoses. Cancer Epidemiol Biomarkers Prev 19(3):716–721

Koskela P, Anttila T, Bjørge T et al (2000) *Chlamydia trachomatis* infection as a risk factor for invasive cervical cancer. Int J Cancer 85(1):35–39

Koutsky LA, Holmes KK, Critchlow CW et al (1992) A cohort study of the risk of cervical intraepithelial neoplasia grade 2 or 3 in relation to papillomavirus infection. N Engl J Med 327(18):1272–1278

Koyi H, Brandén E, Gnarpe J et al (1999) *Chlamydia pneumoniae* may be associated with lung cancer. Preliminary report on a seroepidemiological study. APMIS 107(9):828–832

Koyi H, Brandén E, Gnarpe J et al (2001) An association between chronic infection with *Chlamydia pneumoniae* and lung cancer. A prospective 2-year study. APMIS 109(9):572–580

Kreuzer M, Gerken M, Kreienbrock L et al (2001) Lung cancer in lifetime nonsmoking men – results of a case-control study in Germany. Br J Cancer 84(1):134–140

Kreuzer M, Heinrich J, Kreienbrock L et al (2002) Risk factors for lung cancer among nonsmoking women. Int J Cancer 100(6):706–713

Kuboniwa M, Hasegawa Y, Mao S et al (2008) *P. gingivalis* accelerates gingival epithelial cell progression through the cell cycle. Microbes Infect 10(2):122–128

Kullberg MC, Ward JM, Gorelick PL et al (1998) *Helicobacter hepaticus* triggers colitis in specific-pathogen-free interleukin-10 (IL-10)-deficient mice through an IL-12- and gamma interferon-dependent mechanism. Infect Immun 66(11):5157–5166

Kupferwasser I, Darius H, Müller AM et al (1998) Clinical and morphological characteristics in *Streptococcus bovis* endocarditis: a comparison with other causative microorganisms in 177 cases. Heart 80(3):276–280

Kuramitsu HK, Miyakawa H, Qi M et al (2002) Cellular responses to oral pathogens. Ann Periodontol 7(1):90–94

Kurkivuori J, Salaspuro V, Kaihovaara P et al (2007) Acetaldehyde production from ethanol by oral streptococci. Oral Oncol 43(2):181–186

Kütting B, Bonsmann G, Metze D et al (1997) *Borrelia burgdorferi*-associated primary cutaneous B cell lymphoma: complete clearing of skin lesions after antibiotic pulse therapy or intralesional injection of interferon alfa-2a. J Am Acad Dermatol 36(2 Pt 2):311–314

Kwon Y, Dang ND, Elmunzer BJ (2009) Gastrointestinal stromal tumor complicated by *Streptococcus milleri* bacteremia and liver abscess. Eur J Gastroenterol Hepatol 21(7):824–826

La Vecchia C, Negri E, D'Avanzo B, Savoldelli R, Franceschi S (1991) Genital and urinary tract diseases and bladder cancer. Cancer Res 51(2):629–631

La Vecchia C, Franceschi S, Talamini R et al (1993) Marital status, indicators of sexual activity and prostatic cancer. J Epidemiol Community Health 47(6):450–453

Laurila AL, Anttila T, Läärä E et al (1997) Serological evidence of an association between *Chlamydia pneumoniae* infection and lung cancer. Int J Cancer 74(1):31–34

Le CH, Ko YC, Cheng LS et al (2001) The heterogeneity in risk factors of lung cancer and the difference of histologic distribution between genders in Taiwan. Cancer Causes Control 12(4):289–300

Lecuit M, Abachin E, Martin A et al (2004) Immunoproliferative small intestinal disease associated with *Campylobacter jejuni*. N Engl J Med 350(3):239–248

Lee RA, Woo PC, To AP et al (2003) Geographical difference of disease association in *Streptococcus bovis* bacteraemia. J Med Microbiol 52(Pt 10):903–908

Lees RE, Steele R, Wardle D (1985) Arsenic, syphilis, and cancer of the prostate. J Epidemiol Community Health 39(3):227–230

Legakis NJ, Ioannides H, Tzannetis S et al (1981) Faecal bacterial flora in patients with colon cancer and control subjects. Zentralbl Bakteriol Mikrobiol Hyg A 251(1):54–61

Legier JF (1991) *Streptococcus salivarius* meningitis and colonic carcinoma. South Med J 84(8):1058–1059

Lehtinen M, Dillner J, Knekt P et al (1996) Serologically diagnosed infection with human papillomavirus type 16 and risk for subsequent development of cervical carcinoma: nested case-control study. BMJ 312(7030):537–539

Lehtinen M, Koskela P, Jellum E et al (2002) Herpes simplex virus and risk of cervical cancer: a longitudinal, nested case-control study in the nordic countries. Am J Epidemiol 156(8):687–692

Lehtinen M, Ault KA, Lyytikainen E et al (2011) *Chlamydia trachomatis* infection and risk of cervical intraepithelial neoplasia. Sex Transm Infect 87(5):372–376

Leport C, Bure A, Leport J et al (1987) Incidence of colonic lesions in *Streptococcus bovis* and enterococcal endocarditis. Lancet 1(8535):748

Leung WK, Wu Q, Hannam PM et al (2002) Treponema denticola may stimulate both epithelial proliferation and apoptosis through MAP kinase signal pathways. J Periodontal Res 37(6):445–455

Levy BS, Brooks RJ (1978) More on *Streptococcus bovis* endocarditis and bowel carcinoma. N Engl J Med 298(10):572–573

Li X, Fox JG, Whary MT et al (1998) SCID/NCr mice naturally infected with *Helicobacter hepaticus* develop progressive hepatitis, proliferative typhlitis, and colitis. Infect Immun 66(11):5477–5484

Li C, Inagaki H, Kuo TT et al (2003) Primary cutaneous marginal zone B-cell lymphoma: a molecular and clinicopathologic study of 24 Asian cases. Am J Surg Pathol 27(8):1061–1069

Liang H, Guan P, Yin Z et al (2009) Risk of lung cancer following nonmalignant respiratory conditions among nonsmoking women living in Shenyang, Northeast China. J Womens Health (Larchmt) 18(12):1989–1995

Liddle AD, Abram S, Iyer S (2012) Streptococcus gallolyticus prosthetic joint infection associated with undiagnosed colonic malignancy. Knee Surg Sports Traumatol Arthrosc 20(6):1069–1070

Lightfoot N, Conlon M, Kreiger N et al (2004) Medical history, sexual, and maturational factors and prostate cancer risk. Ann Epidemiol 14(9):655–662

Lin CY, Chao PC, Hong GJ et al (2008) Infective endocarditis from *Streptococcus viridans* associated with colonic carcinoma: a case report. J Card Surg 23(3):263–265

Littman AJ, Thornquist MD, White E et al (2004a) Prior lung disease and risk of lung cancer in a large prospective study. Cancer Causes Control 15(8):819–827

Littman AJ, White E, Jackson LA et al (2004b) *Chlamydia pneumoniae* infection and risk of lung cancer. Cancer Epidemiol Biomarkers Prev 13(10):1624–1630

Littman AJ, Jackson LA, Vaughan TL (2005) *Chlamydia pneumoniae* and lung cancer: epidemiologic evidence. Cancer Epidemiol Biomarkers Prev 14(4):773–778

Liu YC, Ohyashiki JH, Ito Y et al (2006) *Chlamydia psittaci* in ocular adnexal lymphoma: Japanese experience. Leuk Res 30(12):1587–1589

Liyanage NP, Manthey KC, Dassanayake RP et al (2010) *Helicobacter hepaticus* cytolethal distending toxin causes cell death in intestinal epithelial cells via mitochondrial apoptotic pathway. Helicobacter 15(2):98–107

Löfmark S, de Klerk N, Aro H (2011) *Neisseria gonorrhoeae* infection induces altered amphiregulin processing and release. PLoS One 6(1):e16369

López-Boado YS, Wilson CL, Hooper LV et al (2000) Bacterial exposure induces and activates matrilysin in mucosal epithelial cells. J Cell Biol 148(6):1305–1315

Lowenfels AB, Lindström CG, Conway MJ et al (1985) Gallstones and risk of gallbladder cancer. J Natl Cancer Inst 75(1):77–80

Lukic A, Canzio C, Patella A et al (2006) Determination of cervicovaginal microorganisms in women with abnormal cervical cytology: the role of *Ureaplasma urealyticum*. Anticancer Res 26(6C):4843–4849

Lynge E, Jensen OM (1985) Cohort trends in incidence of cervical cancer in Denmark in relation to gonorrheal infection. Acta Obstet Gynecol Scand 64(4):291–296

Ma HM, Shyu KG, Hwang JJ et al (1992) *Streptococcus bovis* endocarditis associated with colonic adenocarcinoma: report of a case. J Formos Med Assoc 91(8):814–817

Macaluso A, Simmang C, Anthony T (1998) *Streptococcus sanguis* bacteremia and colorectal cancer. South Med J 91(2):206–207

Madeleine MM, Anttila T, Schwartz SM et al (2007) Risk of cervical cancer associated with *Chlamydia trachomatis* antibodies by histology, HPV type and HPV cofactors. Int J Cancer 120(3):650–655

Mager DL, Haffajee AD, Devlin PM et al (2005) The salivary microbiota as a diagnostic indicator of oral cancer: a descriptive, non-randomized study of cancer-free and oral squamous cell carcinoma subjects. J Transl Med 3:27

Maggio-Price L, Shows D, Waggie K et al (2002) *Helicobacter bilis* infection accelerates and H. hepaticus infection delays the development of colitis in multiple drug resistance-deficient (mdr1a–/–) mice. Am J Pathol 160(2):739–751

Maggio-Price L, Bielefeldt-Ohmann H, Treuting P et al (2005) Dual infection with *Helicobacter bilis* and *Helicobacter hepaticus* in p-glycoprotein-deficient mdr1a –/– mice results in colitis that progresses to dysplasia. Am J Pathol 166(6):1793–1806

Maggio-Price L, Treuting P, Zeng W et al (2006) *Helicobacter* infection is required for inflammation and colon cancer in SMAD3-deficient mice. Cancer Res 66(2):828–838

Mann GN, Scoggins CR, Adkins B (1997) Perforated cecal adenocarcinoma presenting as a thigh abscess. South Med J 90(9):949–951

Mao S, Park Y, Hasegawa Y et al (2007) Intrinsic apoptotic pathways of gingival epithelial cells modulated by *Porphyromonas gingivalis*. Cell Microbiol 9(8):1997–2007

Markowska J, Fischer N, Filas V et al (1999) The role of *Chlamydia trachomatis* infection in cervical cancer development. Eur J Gynaecol Oncol 20(2):144–146

Matsukura N, Yokomuro S, Yamada S et al (2002) Association between *Helicobacter bilis* in bile and biliary tract malignancies: *H. bilis* in bile from Japanese and Thai patients with benign and malignant diseases in the biliary tract. Jpn J Cancer Res 93(7):842–847

Matthews JM, Moreno LI, Dennis J et al (2008) Ocular adnexal lymphoma: no evidence for bacterial DNA associated with lymphoma pathogenesis. Br J Haematol 142(2):246–249

Maurer KJ, Ihrig MM, Rogers AB et al (2005) Identification of cholelithogenic enterohepatic *Helicobacter* species and their role in murine cholesterol gallstone formation. Gastroenterology 128(4):1023–1033

Maurya SP (2011) *Streptococcus bovis* bacteremia and colorectal carcinoma. Indian J Cancer 48(3):375–376

McMahon AJ, Auld CD, Dale BA et al (1991) *Streptococcus bovis* septicaemia associated with uncomplicated colonic carcinoma. Br J Surg 78(7):883–885

Michalek AM, Mahoney MC, McLaughlin CC et al (1994) Historical and contemporary correlates of syphilis and cancer. Int J Epidemiol 23(2):381–385

Michaud DS, Platz EA, Giovannucci E (2007) Gonorrhoea and male bladder cancer in a prospective study. Br J Cancer 96(1):169–171

Mihara J, Miyazawa Y, Holt SC (1993) Modulation of growth and function of human gingival fibroblasts by fibroblast-activating factor derived from *Porphyromonas gingivalis* W50. Infect Immun 61(2):596–601

Mikamo H, Izumi K, Ito K et al (1993) Internal bacterial flora of solid uterine cervical cancer. Kansenshogaku Zasshi 67(11):1057–1061

Mikamo H, Sato Y, Hayasaki Y et al (1999) Intravaginal bacterial flora in patients with uterine cervical cancer. High incidence of detection of *Gardnerella vaginalis*. J Infect Chemother 5(2):82–85

Millichap JJ, McKendrick AI, Drelichman VS (2005) *Streptococcus intermedius* liver abscesses and colon cancer: a case report. West Indian Med J 54(5):341–342

Mishina T, Watanabe H, Araki H et al (1985) Epidemiological study of prostatic cancer by matched-pair analysis. Prostate 6(4):423–436

Mizutani T, Mitsuoka T (1979) Effect of intestinal bacteria on incidence of liver tumors in gnotobiotic C3H/He male mice. J Natl Cancer Inst 63(6):1365–1370

Mizutani T, Mitsuoka T (1980) Inhibitory effect of some intestinal bacteria on liver tumorigenesis in gnotobiotic C3H/He male mice. Cancer Lett 11(2):89–95

Mohm J, Naumann R, Schuler U et al (1998) Abdominal lymphomas, convulsive seizure and coma: a case of successfully treated, advanced Whipple's disease with cerebral involvement. Eur J Gastroenterol Hepatol 10(10):893–895

Mommsen S, Sell A (1983) Prostatic hypertrophy and venereal disease as possible risk factors in the development of bladder cancer. Urol Res 11(2):49–52

Monari P, Farisoglio C, Calzavara Pinton PG et al (2007) *Borrelia burgdorferi*-associated primary cutaneous marginal-zone B-cell lymphoma: a case report. Dermatology 215(3):229–232

Montalto M, Izzi D, La Regina M et al (2004) Non-life-threatening sepsis: report of two cases. Am J Med Sci 327(5):275–277

Moore WE, Moore LH (1995) Intestinal floras of populations that have a high risk of colon cancer. Appl Environ Microbiol 61(9):3202–3207

Morita E, Narikiyo M, Yano A et al (2003) Different frequencies of *Streptococcus anginosus* infection in oral cancer and esophageal cancer. Cancer Sci 94(6):492–496

Morita E, Narikiyo M, Yokoyama A et al (2005) Predominant presence of *Streptococcus anginosus* in the saliva of alcoholics. Oral Microbiol Immunol 20(6):362–365

Moutel K, Merol JC, Prevost A et al (2004) Prevalence of *Chlamydia pneumoniae* infection in squamous cell carcinoma of the head and neck. Anticancer Res 24(2C):1207–1210

Mulder MM, Heddema ER, Pannekoek Y et al (2006) No evidence for an association of ocular adnexal lymphoma with *Chlamydia psittaci* in a cohort of patients from the Netherlands. Leuk Res 30(10):1305–1307

Munksgaard L, Frisch M, Melbye M et al (2000) Incidence patterns of lyme disease and cutaneous B-cell non-Hodgkin's lymphoma in the United States. Dermatology 201(4):351–352

Muñoz N, Bosch FX, de Sanjosé S et al (1993) Risk factors for cervical intraepithelial neoplasia grade III/carcinoma in situ in Spain and Colombia. Cancer Epidemiol Biomarkers Prev 2(5):423–431

Muñoz N, Castellsagué X, Bosch FX et al (1996) Difficulty in elucidating the male role in cervical cancer in Colombia, a high-risk area for the disease. J Natl Cancer Inst 88(15):1068–1075

Murata H, Tsuji S, Tsujii M et al (2004) *Helicobacter bilis* infection in biliary tract cancer. Aliment Pharmacol Ther 20(Suppl 1):90–94

Murray HW, Roberts RB (1978) *Streptococcus bovis* bacteremia and underlying gastrointestinal disease. Arch Intern Med 138(7):1097–1099

Nagamine CM, Rogers AB, Fox JG et al (2008a) *Helicobacter hepaticus* promotes azoxymethane-initiated colon tumorigenesis in BALB/c-IL10-deficient mice. Int J Cancer 122(4):832–838

Nagamine CM, Sohn JJ, Rickman BH et al (2008b) *Helicobacter hepaticus* infection promotes colon tumorigenesis in the BALB/c-Rag2(−/−) Apc(Min/+) mouse. Infect Immun 76(6):2758–2766

Nakamura J, Kubota Y, Miyaoka M et al (2002) Comparison of four microbial enzymes in Clostridia and Bacteroides isolated from human feces. Microbiol Immunol 46(7):487–490

Nakhjiri SF, Park Y, Yilmaz O et al (2001) Inhibition of epithelial cell apoptosis by *Porphyromonas gingivalis*. FEMS Microbiol Lett 200(2):145–149

Nalbandian A, Yan BS, Pichugin A et al (2009) Lung carcinogenesis induced by chronic tuberculosis infection: the experimental model and genetic control. Oncogene 28(17):1928–1938

Narikiyo M, Tanabe C, Yamada Y et al (2004) Frequent and preferential infection of *Treponema denticola*, *Streptococcus mitis*, and *Streptococcus anginosus* in esophageal cancers. Cancer Sci 95(7):569–574

Nath G, Singh H, Shukla VK (1997) Chronic typhoid carriage and carcinoma of the gallbladder. Eur J Cancer Prev 6(6):557–559

Nath G, Singh YK, Kumar K et al (2008) Association of carcinoma of the gallbladder with typhoid carriage in a typhoid endemic area using nested PCR. J Infect Dev Ctries 2(4):302–307

Naucler P, Chen HC, Persson K et al (2007) Seroprevalence of human papillomaviruses and *Chlamydia trachomatis* and cervical cancer risk: nested case-control study. J Gen Virol 88 (Pt 3):814–822

Ness RB, Goodman MT, Shen C et al (2003) Serologic evidence of past infection with *Chlamydia trachomatis*, in relation to ovarian cancer. J Infect Dis 187(7):1147–1152

Ness RB, Shen C, Bass D et al (2008) *Chlamydia trachomatis* serology in women with and without ovarian cancer. Infect Dis Obstet Gynecol 2008:219672

Neuer A, Menton M (1995) Bacteriological findings in patients with cervical intra-epithelial neoplasia. Zentralbl Gynakol 117(8):435–438

Nijjer S, Dubrey SW (2010) *Streptococcus sanguis* endocarditis associated with colonic carcinoma. BMJ Case Rep pii:bcr09.2009.2311

Nilsson HO, Stenram U, Ihse I et al (2006) *Helicobacter* species ribosomal DNA in the pancreas, stomach and duodenum of pancreatic cancer patients. World J Gastroenterol 2(19):3038–3043

Noble CJ, Uttley AH, Falk RH et al (1978) *Streptococcus bovis* endocarditis and colonic cancer. Lancet 1(8067):766

Noguchi N, Ohashi T, Shiratori T et al (2007) Association of tannase-producing *Staphylococcus lugdunensis* with colon cancer and characterization of a novel tannase gene. J Gastroenterol 42(5):346–351

Norfleet RG, Mitchell PD (1993) Streptococcus bovis does not selectively colonize colorectal cancer and polyps. J Clin Gastroenterol 17(1):25–28

Olejek A, Kozak-Darmas I, Kellas-Sleczka S et al (2009) *Chlamydia trachomatis* infection in women with lichen sclerosus vulvae and vulvar cancer. Neuro Endocrinol Lett 30(5):671–674

Oliveira Mde L, Amorim MM, Souza AS et al (2008) Chlamydia infection in patients with and without cervical intra-epithelial lesions. Rev Assoc Med Bras 54(6):506–512

Osann KE, Lowery JT, Schell MJ (2000) Small cell lung cancer in women: risk associated with smoking, prior respiratory disease, and occupation. Lung Cancer 28(1):1–10

Paavonen J, Vesterinen E, Meyer B et al (1979) Genital *Chlamydia trachomatis* infections in patients with cervical atypia. Obstet Gynecol 54(3):289–291

Paavonen J, Karunakaran KP, Noguchi Y et al (2003) Serum antibody response to the heat shock protein 60 of *Chlamydia trachomatis* in women with developing cervical cancer. Am J Obstet Gynecol 189(5):1287–1292

Pandey M, Shukla VK (2003) Lifestyle, parity, menstrual and reproductive factors and risk of gallbladder cancer. Eur J Cancer Prev 12(4):269–272

Pandey M, Mishra RR, Dixit R et al (2010) *Helicobacter bilis* in human gallbladder cancer: results of a case-control study and a meta-analysis. Asian Pac J Cancer Prev 11(2):343–347

Pelucchi C, Talamini R, Negri E et al (2006) Genital and urinary tract diseases and prostate cancer risk. Eur J Cancer Prev 15(3):254–257

Pergola V, Di Salvo G, Habib G et al (2001) Comparison of clinical and echocardiographic characteristics of *Streptococcus bovis* endocarditis with that caused by other pathogens. Am J Cardiol 88(8):871–875

Peters BS, Thomas BJ, Yi C et al (1995) *Chlamydia trachomatis* as a possible cofactor for Kaposi's sarcoma in AIDS. Int J STD AIDS 6(6):422–425

Pigrau C, Lorente A, Pahissa A et al (1988) *Streptococcus bovis* bacteremia and digestive system neoplasms. Scand J Infect Dis 20(4):459–460

Ponzoni M, Ferreri AJ, Guidoboni M et al (2008) Chlamydia infection and lymphomas: association beyond ocular adnexal lymphomas highlighted by multiple detection methods. Clin Cancer Res 14(18):5794–5800

Ponzoni M, Bonetti F, Poliani PL et al (2011a) Central nervous system marginal zone B-cell lymphoma associated with *Chlamydophila psittaci* infection. Hum Pathol 42(5):738–742

Ponzoni M, Ferreri AJ, Mappa S et al (2011b) Prevalence of *Borrelia burgdorferi* infection in a series of 98 primary cutaneous lymphomas. Oncologist 16(11):1582–1588

Potter MA, Cunliffe NA, Smith M, Miles RS, Flapan AD, Dunlop MG (1998) A prospective controlled study of the association of Streptococcus bovis with colorectal carcinoma. J Clin Pathol 51(6):473–474

Pradhan SB, Dali S (2004) Relation between gallbladder neoplasm and *Helicobacter hepaticus* infection. Kathmandu Univ Med J (KUMJ) 2(4):331–335

Prozialeck WC, Fay MJ, Lamar PC et al (2002) *Chlamydia trachomatis* disrupts N-cadherin-dependent cell-cell junctions and sequesters beta-catenin in human cervical epithelial cells. Infect Immun 70(5):2605–2613

Putnins EE, Sanaie AR, Wu Q et al (2002) Induction of keratinocyte growth factor 1 expression by lipopolysaccharide is regulated by CD-14 and toll-like receptors 2 and 4. Infect Immun 70(12):6541–6548

Quint KD, de Koning MN, Geraets DT et al (2009) Comprehensive analysis of Human papillomavirus and *Chlamydia trachomatis* in in-situ and invasive cervical adenocarcinoma. Gynecol Oncol 114(3):390–394

Qvigstad E, Onsrud M, Degré M (1985) Cell-mediated and humoral immune responses to chlamydial and herpesvirus antigens in patients with cervical carcinoma. Gynecol Oncol 20(2):184–189

Ramanakumar AV, Parent ME, Menzies D et al (2006) Risk of lung cancer following nonmalignant respiratory conditions: evidence from two case-control studies in Montreal, Canada. Lung Cancer 53(1):5–12

Ramljak D, Jones AB, Diwan BA et al (2000) Epidermal growth factor and transforming growth factor-alpha-associated overexpression of cyclin D1, Cdk4, and c-Myc during hepatocarcinogenesis in *Helicobacter hepaticus*-infected A/JCr mice. Cancer Res 58(16):3590–3597

Randi G, Franceschi S, La Vecchia C (2006) Gallbladder cancer worldwide: geographical distribution and risk factors. Int J Cancer 118(7):1591–1602

Rao VP, Poutahidis T, Ge Z et al (2006) Innate immune inflammatory response against enteric bacteria *Helicobacter hepaticus* induces mammary adenocarcinoma in mice. Cancer Res 66(15):7395–7400

Reesink-Peters N, Ossewaarde JM, Van Der Zee AG et al (2001) No association of anti-*Chlamydia trachomatis* antibodies and severity of cervical neoplasia. Sex Transm Infect 77(2):101–102

Resto-Ruiz SI, Schmiederer M, Sweger D et al (2002) Induction of a potential paracrine angiogenic loop between human THP-1 macrophages and human microvascular endothelial cells during *Bartonella henselae* infection. Infect Immun 70(8):4564–4570

Reynolds JG, Silva E, McCormack WM (1983) Association of *Streptococcus bovis* bacteremia with bowel disease. J Clin Microbiol 17(4):696–697

Riska A, Finne P, Alfthan H et al (2006) Past chlamydial infection is not associated with primary fallopian tube carcinoma. Eur J Cancer 42(12):1835–1838

Robbins N, Klein RS (1983) Carcinoma of the colon 2 years after endocarditis due to *Streptococcus bovis*. Am J Gastroenterol 78(3):162–163

Robbins S, Chuang VP, Hersh T (1988) The development of hepatobiliary cancer in a carrier of *Salmonella typhus*. Am J Gastroenterol 83(6):675–678

Rogers AB, Boutin SR, Whary MT et al (2004) Progression of chronic hepatitis and preneoplasia in *Helicobacter hepaticus*-infected A/JCr mice. Toxicol Pathol 32(6):668–677

Roggero E, Zucca E, Mainetti C et al (2000) Eradication of *Borrelia burgdorferi* infection in primary marginal zone B-cell lymphoma of the skin. Hum Pathol 31(2):263–268

Rosado MF, Byrne GE Jr, Ding F et al (2006) Ocular adnexal lymphoma: a clinicopathologic study of a large cohort of patients with no evidence for an association with *Chlamydia psittaci*. Blood 107(2):467–472

Rosenblatt KA, Wicklund KG, Stanford JL (2001) Sexual factors and the risk of prostate cancer. Am J Epidemiol 153(12):1152–1158

Rössler MJ, Rappl G, Muche M et al (2003) No evidence of skin infection with *Chlamydia pneumoniae* in patients with cutaneous T cell lymphoma. Clin Microbiol Infect 9(7):721–723

Ruiz A, Reischl U, Swerdlow SH et al (2007) Extranodal marginal zone B-cell lymphomas of the ocular adnexa: multiparameter analysis of 34 cases including interphase molecular cytogenetics and PCR for *Chlamydia psittaci*. Am J Surg Pathol 31(5):792–802

Ruoff KL, Miller SI, Garner CV et al (1989) Bacteremia with *Streptococcus bovis* and *Streptococcus salivarius*: clinical correlates of more accurate identification of isolates. J Clin Microbiol 27(2):305–308

Safaeian M, Quint K, Schiffman M et al (2010) *Chlamydia trachomatis* and risk of prevalent and incident cervical premalignancy in a population-based cohort. J Natl Cancer Inst 102(23):1794–1804

Safaeian M, Gao YT, Sakoda LC et al (2011) Chronic typhoid infection and the risk of biliary tract cancer and stones in Shanghai, China. Infect Agent Cancer 6:6

Samet JM, Humble CG, Pathak DR (1986) Personal and family history of respiratory disease and lung cancer risk. Am Rev Respir Dis 134(3):466–470

Sarma AV, McLaughlin JC, Wallner LP et al (2006) Sexual behavior, sexually transmitted diseases and prostatitis: the risk of prostate cancer in black men. J Urol 176(3):1108–1113

Sasaki H, Igaki H, Ishizuka T et al (1995) Presence of *Streptococcus* DNA sequence in surgical specimens of gastric cancer. Jpn J Cancer Res 86(9):791–794

Sasaki H, Ishizuka T, Muto M et al (1998) Presence of *Streptococcus anginosus* DNA in esophageal cancer, dysplasia of esophagus, and gastric cancer. Cancer Res 58(14):2991–2995

Sasaki M, Yamaura C, Ohara-Nemoto Y et al (2005) *Streptococcus anginosus* infection in oral cancer and its infection route. Oral Dis 11(3):151–156

Schabath MB, Delclos GL, Martynowicz MM et al (2005) Opposing effects of emphysema, hay fever, and select genetic variants on lung cancer risk. Am J Epidemiol 161(5):412–422

Schachter J, Hill EC, King EB et al (1975) Chlamydial infection in women with cervical dysplasia. Am J Obstet Gynecol 123(7):753–757

Schachter J, Hill EC, King EB et al (1982) *Chlamydia trachomatis* and cervical neoplasia. JAMA 248(17):2134–2138

Schöllkopf C, Melbye M, Munksgaard L et al (2008) Borrelia infection and risk of non-Hodgkin lymphoma. Blood 111(12):5524–5529

Schwartz AG, Yang P, Swanson GM (1996) Familial risk of lung cancer among nonsmokers and their relatives. Am J Epidemiol 144(6):554–562

Seglenieks A, Black RB (1998) *Streptococcus bovis* and its association with bowel cancer. Aust N Z J Surg 68(7):542–543

Seifert H (2009) The clinical importance of microbiological findings in the diagnosis and management of bloodstream infections. Clin Infect Dis 48(Suppl 4):S238–S245

Serra I, Yamamoto M, Calvo A et al (2002) Association of chili pepper consumption, low socioeconomic status and longstanding gallstones with gallbladder cancer in a Chilean population. Int J Cancer 102(4):407–411

Severi G, Shannon BA, Hoang HN et al (2010) Plasma concentration of *Propionibacterium acnes* antibodies and prostate cancer risk: results from an Australian population-based case-control study. Br J Cancer 103(3):411–415

Shanmughapriya S, Senthilkumar G, Vinodhini K et al (2012) Viral and bacterial aetiologies of epithelial ovarian cancer. Eur J Clin Microbiol Infect Dis 31(9):2311–2317

Shannon BA, Cohen RJ, Garrett KL (2006) Polymerase chain reaction-based identification of *Propionibacterium acnes* types isolated from the male urinary tract: evaluation of adolescents, normal adults and men with prostatic pathology. BJU Int 98(2):388–392

Shannon BA, Cohen RJ, Garrett KL (2008) The antibody response to *Propionibacterium acnes* is an independent predictor of serum prostate-specific antigen levels in biopsy-negative men. BJU Int 101(4):429–435

Sheehan-Dare RA, Cunliffe WJ, Simmons AV et al (1988) Acne vulgaris and malignancy. Br J Dermatol 119(5):669–673

Shen D, Yuen HK, Galita DA et al (2006) Detection of *Chlamydia pneumoniae* in a bilateral orbital mucosa-associated lymphoid tissue lymphoma. Am J Ophthalmol 141(6):1162–1163

Shen XJ, Rawls JF, Randall T et al (2010) Molecular characterization of mucosal adherent bacteria and associations with colorectal adenomas. Gut Microbes 1(3):138–147

Sherman KJ, Daling JR, Chu J et al (1991) Genital warts, other sexually transmitted diseases, and vulvar cancer. Epidemiology 2(4):257–262

Sheth S, Bedford A, Chopra S (2000) Primary gallbladder cancer: recognition of risk factors and the role of prophylactic cholecystectomy. Am J Gastroenterol 95(6):1402–1410

Shiga K, Tateda M, Saijo S et al (2001) Presence of Streptococcus infection in extra-oropharyngeal head and neck squamous cell carcinoma and its implication in carcinogenesis. Oncol Rep 8(2):245–248

Shimoyama T, Takahashi R, Abe D et al (2010) Serological analysis of *Helicobacter hepaticus* infection in patients with biliary and pancreatic diseases. J Gastroenterol Hepatol 25(suppl 1):S86–S89

Shukla VK, Singh H, Pandey M et al (2000) Carcinoma of the gallbladder – is it a sequel of typhoid? Dig Dis Sci 45(5):900–903

Siegert CE, Overbosch D (1995) Carcinoma of the colon presenting as *Streptococcus sanguis* bacteremia. Am J Gastroenterol 90(9):1528–1529

Silver SC (1984) *Streptococcus bovis* endocarditis and its association with colonic carcinoma. Dis Colon Rectum 27(9):613–614

Singh H, Pandey M, Shukla VK (1996) Salmonella carrier state, chronic bacterial infection and gallbladder carcinogenesis. Eur J Cancer Prev 5(2):144

Sipowicz MA, Weghorst CM, Shiao YH et al (1997) Lack of p53 and ras mutations in *Helicobacter hepaticus*-induced liver tumors in A/JCr mice. Carcinogenesis 18(1):233–236

Smit EF, Wijnja L, Manson WL (1994) Bacteremia caused by *Streptococcus salivarius* and *S. milleri*, and colonic carcinoma. Ned Tijdschr Geneeskd 138(30):1529–1531

Smith TR, Friedman SN, Adler H (1978) *Streptococcus bovis* septicemia as a clue to colon neoplasms in two cases. Am J Roentgenol 131(5):887–888

Smith JS, Muñoz N, Franceschi S, Eluf-Neto J, Herrero R, Peeling RW (2001) Chlamydia trachomatis and cervical squamous cell carcinoma. JAMA 285(13):1704

Smith JS, Muñoz N, Herrero R et al (2002) Evidence for *Chlamydia trachomatis* as a human papillomavirus cofactor in the etiology of invasive cervical cancer in Brazil and the Philippines. J Infect Dis 185(3):324–331

Smith JS, Bosetti C, Muñoz N et al (2004) *Chlamydia trachomatis* and invasive cervical cancer: a pooled analysis of the IARC multicentric case-control study. Int J Cancer 111(3):431–439

Smith JS, Kumlin U, Nyberg F et al (2008) Lack of association between serum antibodies of *Chlamydia pneumoniae* infection and the risk of lung cancer. Int J Cancer 123(10):2469–2471

Sobhani I, Tap J, Roudot-Thoraval F et al (2011) Microbial dysbiosis in colorectal cancer (CRC) patients. PLoS One 6(1):e16393

Srivastava A, Walter N, Atkinson P (2010) *Streptococcus bovis* infection of total hip arthroplasty in association with carcinoma of colon. J Surg Orthop Adv 19(2):125–128

Steinberg D, Naggar CZ (1977) *Streptococcus bovis* endocarditis with carcinoma of the colon. N Engl J Med 297(24):1354–1355

Sterzenbach T, Bartonickova L, Behrens W et al (2008) Role of the *Helicobacter hepaticus* flagellar sigma factor FliA in gene regulation and murine colonization. J Bacteriol 190(19):6398–6408

Stojan T, Müller W, Pfyffer G (2006) *Streptococcus milleri.* An early sign of cancer in otorhinolaryngologic patients? HNO 54(10):778–780

Stone KM, Zaidi A, Rosero-Bixby L et al (1995) Sexual behavior, sexually transmitted diseases, and risk of cervical cancer. Epidemiology 6(4):409–414

Strom BL, Soloway RD, Rios-Dalenz JL et al (1995) Risk factors for gallbladder cancer. An international collaborative case-control study. Cancer 76(10):1747–1756

Sugano N, Yokoyama K, Oshikawa M et al (2003) Detection of *Streptococcus anginosus* and 8-hydroxydeoxyguanosine in saliva. J Oral Sci 45(4):181–184

Sutcliffe S, Giovannucci E, De Marzo AM et al (2006) Gonorrhea, syphilis, clinical prostatitis, and the risk of prostate cancer. Cancer Epidemiol Biomarkers Prev 15(11):2160–2166

Sutcliffe S, Giovannucci E, Gaydos CA et al (2007a) Plasma antibodies against *Chlamydia trachomatis*, human papillomavirus, and human herpesvirus type 8 in relation to prostate cancer: a prospective study. Cancer Epidemiol Biomarkers Prev 16(8):1573–1580

Sutcliffe S, Giovannucci E, Isaacs WB et al (2007b) Acne and risk of prostate cancer. Int J Cancer 121(12):2688–2692

Tabibian N, Clarridge JE (1989) *Streptococcus bovis* septicemia and large bowel neoplasia. Am Fam Physician 39(4):227–229

Takac I, Gorisek B (1998) The frequency of *Chlamydia trachomatis* infection in women with intraepithelial neoplasia of the uterine cervix. Eur J Gynaecol Oncol 19(5):492–494

Takac I, Gorisek B (1999) *Chlamydia trachomatis* infection in women with and without cervical intraepithelial neoplasia. Int J STD AIDS 10(5):331–333

Takayama S, Takahashi H, Matsuo Y et al (2010) Effect of *Helicobacter bilis* infection on human bile duct cancer cells. Dig Dis Sci 55(7):1905–1910

Takemura A, Matsuda N, Kimura S et al (1998) *Porphyromonas gingivalis* lipopolysaccharide modulates the responsiveness of human periodontal ligament fibroblasts to platelet-derived growth factor. J Periodontal Res 33(7):400–407

Takeyama K, Mitsuzawa H, Shimizu T et al (2006) Prostate cell lines secrete IL-8 in response to *Mycoplasma hominis* through Toll-like receptor 2-mediated mechanism. Prostate 66(4):386–391

Takino H, Li C, Hu S et al (2008) Primary cutaneous marginal zone B-cell lymphoma: a molecular and clinicopathological study of cases from Asia, Germany, and the United States. Mod Pathol 21(12):1517–1526

Tateda M, Shiga K, Saijo S et al (2000) *Streptococcus anginosus* in head and neck squamous cell carcinoma: implication in carcinogenesis. Int J Mol Med 6(6):699–703

Tavani A, Rosato V, Di Palma F et al (2012) History of cholelithiasis and cancer risk in a network of case-control studies. Ann Oncol 23(8):2173–2178

Taylor ML, Mainous AG 3rd, Wells BJ (2005) Prostate cancer and sexually transmitted diseases: a meta-analysis. Fam Med 37(7):506–512

Tewari M, Mishra RR, Shukla HS (2010) *Salmonella typhi* and gallbladder cancer: report from an endemic region. Hepatobiliary Pancreat Dis Int 9(5):524–530

Theve EJ, Feng Y, Taghizadeh K et al (2008) Sex hormone influence on hepatitis in young male A/JCr mice infected with *Helicobacter hepaticus*. Infect Immun 76(9):4071–4078

Tjalsma H, Schöller-Guinard M, Lasonder E et al (2006) Profiling the humoral immune response in colon cancer patients: diagnostic antigens from *Streptococcus bovis*. Int J Cancer 119(9):2127–2135

Toprak NU, Yagci A, Gulluoglu BM et al (2006) A possible role of *Bacteroides fragilis* enterotoxin in the aetiology of colorectal cancer. Clin Microbiol Infect 12(8):782–786

Trajber I, Solomon A, Michowitz M et al (1984) *Streptococcus bovis* subacute bacterial endocarditis as a presenting symptom of occult double carcinoma of the colon. J Surg Oncol 27(3):186–188

Tripodi MF, Adinolfi LE, Ragone E et al (2004) *Streptococcus bovis* endocarditis and its association with chronic liver disease: an underestimated risk factor. Clin Infect Dis 38(10):1394–1400

Tseng HF, Morgenstern H, Mack TM et al (2003) Risk factors for anal cancer: results of a population-based case – control study. Cancer Causes Control 14(9):837–846

Tzur T, Liberman S, Felzenstein I et al (2003) Liver abscesses caused by *Streptococcus milleri*: an uncommon presenting sign of silent colonic cancer. Isr Med Assoc J 5(3):206–207

Uldry PA, Regli F, Ferrazzini M (1988) Primary lymphoma of the nervous system following radiculoneuritis caused by *Borrelia burgdorferi*: study of a case. Schweiz Arch Neurol Psychiatr 139(1):31–38

Valadan M, Yarandi F, Eftekhar Z et al (2010) *Chlamydia trachomatis* and cervical intraepithelial neoplasia in married women in a Middle Eastern community. East Mediterr Health J 16(3):304–307

Van Tassel RL, MacDonald DK, Wilkins TD (1982) Production of a fecal mutagen by *Bacteroides* spp. Infect Immun 37(3):975–980

Vargas RL, Fallone E, Felgar RE et al (2006) Is there an association between ocular adnexal lymphoma and infection with *Chlamydia psittaci*? The University of Rochester experience. Leuk Res 30(5):547–551

Vaska VL, Faoagali JL (2009) *Streptococcus bovis* bacteraemia: identification within organism complex and association with endocarditis and colonic malignancy. Pathology 41(2):183–186

Vince KG, Kantor SR, Descalzi J (2003) Late infection of a total knee arthroplasty with *Streptococcus bovis* in association with carcinoma of the large intestine. J Arthroplasty 18(6):813–815

Waisberg J, Matheus Cde O, Pimenta J (2002) Infectious endocarditis from *Streptococcus bovis* associated with colonic carcinoma: case report and literature review. Arq Gastroenterol 39(3):177–180

Wallin KL, Wiklund F, Luostarinen T et al (2002) A population-based prospective study of *Chlamydia trachomatis* infection and cervical carcinoma. Int J Cancer 101(4):371–374

Wang SY, Hu YL, Wu YL et al (1996a) A comparative study of the risk factors for lung cancer in Guangdong, China. Lung Cancer 14(Suppl 1):S99–S105

Wang TJ, Zhou BS, Shi JP (1996b) Lung cancer in nonsmoking Chinese women: a case-control study. Lung Cancer 14(suppl 1):S93–S98

Wang S, Ernst LM, Smith BR et al (2003) Systemic *Tropheryma whippleii* infection associated with monoclonal B-cell proliferation: a *Helicobacter pylori*-type pathogenesis? Arch Pathol Lab Med 127(12):1619–1622

Wang XR, Yu IT, Chiu YL et al (2009) Previous pulmonary disease and family cancer history increase the risk of lung cancer among Hong Kong women. Cancer Causes Control 20(5):757–763

Wang T, Cai G, Qiu Y et al (2012) Structural segregation of gut microbiota between colorectal cancer patients and healthy volunteers. ISME J 6(2):320–329

Ward JM, Anver MR, Haines DC et al (1994a) Chronic active hepatitis in mice caused by *Helicobacter hepaticus*. Am J Pathol 145(4):959–968

Ward JM, Fox JG, Anver MR et al (1994b) Chronic active hepatitis and associated liver tumors in mice caused by a persistent bacterial infection with a novel *Helicobacter* species. J Natl Cancer Inst 86(16):1222–1227

Ward JM, Anver MR, Haines DC et al (1996a) Inflammatory large bowel disease in immunodeficient mice naturally infected with *Helicobacter hepaticus*. Lab Anim Sci 46(1):15–20

Ward JM, Benveniste RE, Fox CH et al (1996b) Autoimmunity in chronic active *Helicobacter hepatitis* of mice. Serum antibodies and expression of heat shock protein 70 in liver. Am J Pathol 148(2):509–517

Watanabe K, Rao VP, Poutahidis T et al (2008) Cytotoxic-T-lymphocyte-associated antigen 4 blockade abrogates protection by regulatory T cells in a mouse model of microbially induced innate immune-driven colitis. Infect Immun 76(12):5834–5842

Weitberg AB, Annese C, Ginsberg MB (1981) *Streptococcus bovis* meningitis and carcinoma of the colon. Johns Hopkins Med J 148(6):260–261

Welton JC, Marr JS, Friedman SM (1979) Association between hepatobiliary cancer and typhoid carrier state. Lancet 1(8120):791–794

Weng SW, Liu JW, Chen WJ et al (2005) Recurrent *Klebsiella pneumoniae* liver abscess in a diabetic patient followed by *Streptococcus bovis* endocarditis – occult colon tumor plays an important role. Jpn J Infect Dis 58(2):70–72

Wentling GK, Metzger PP, Dozois EJ et al (2006) Unusual bacterial infections and colorectal carcinoma – *Streptococcus bovis* and *Clostridium septicum*: report of three cases. Dis Colon Rectum 49(8):1223–1227

Wheeler LA, Soderberg FB, Goldman P (1975) The reduction of N-hydroxy-4-acetylaminobiphenyl by the intestinal microflora of the rat. Cancer Res 35(11 Pt 1):2962–2968

Williams MA, Kenya PR, Mati JK et al (1994) Risk factors for invasive cervical cancer in Kenyan women. Int J Epidemiol 23(5):906–912

Wilson WR, Thompson RL, Wilkowske CJ et al (1981) Short-term therapy for streptococcal infective endocarditis. Combined intramuscular administration of penicillin and streptomycin. JAMA 245(4):360–363

Wong A, Maclean AB, Furrows SJ (2007) Could epithelial ovarian cancer be associated with chlamydial infection? Eur J Gynaecol Oncol 28(2):117–120

Wood GS, Kamath NV, Guitart J et al (2001) Absence of *Borrelia burgdorferi* DNA in cutaneous B-cell lymphomas from the United States. J Cutan Pathol 28(10):502–507

Woodworth HL, McCaskey SJ, Duriancik DM et al (2010) Dietary fish oil alters T lymphocyte cell populations and exacerbates disease in a mouse model of inflammatory colitis. Cancer Res 70(20):7960–7969

Wu AH, Yu MC, Thomas DC et al (1988) Personal and family history of lung disease as risk factors for adenocarcinoma of the lung. Cancer Res 48(24 Pt 1):7279–7284

Wu AH, Fontham ET, Reynolds P et al (1995) Previous lung disease and risk of lung cancer among lifetime nonsmoking women in the United States. Am J Epidemiol 141(11):1023–1032

Wu S, Rhee KJ, Albesiano E et al (2009) A human colonic commensal promotes colon tumorigenesis via activation of T helper type 17 Tcell responses. Nat Med 15(9):1016–1022

Wu-Williams AH, Dai XD, Blot W et al (1990) Lung cancer among women in north-east China. Br J Cancer 62(6):982–987

Xu YK, Zhang FL, Feng T et al (2009) Meta-analysis on the correlation of cholecystectomy or cholecystolithiasis to risk of colorectal cancer in Chinese population. Ai Zheng 28(7):749–755

Yagyu K, Lin Y, Obata Y et al (2004) Bowel movement frequency, medical history and the risk of gallbladder cancer death: a cohort study in Japan. Cancer Sci 95(8):674–678

Yakushijin Y, Kodama T, Takaoka I et al (2007) Absence of chlamydial infection in Japanese patients with ocular adnexal lymphoma of mucosa-associated lymphoid tissue. Int J Hematol 85(3):223–230

Yanagisawa N, Geironson L, Al-Soud WA et al (2005) Expression of matrix metalloprotease-2, -7 and -9 on human colon, liver and bile duct cell lines by enteric and gastric *Helicobacter* species. FEMS Immunol Med Microbiol 44(2):197–204

Yang J, Hooper WC, Phillips DJ et al (2003) Induction of proinflammatory cytokines in human lung epithelial cells during *Chlamydia pneumoniae* infection. Infect Immun 71(2):614–620

Yeung L, Tsao YP, Chen PY et al (2004) Combination of adult inclusion conjunctivitis and mucosa-associated lymphoid tissue (MALT) lymphoma in a young adult. Cornea 23(1):71–75

Yilmaz O, Verbeke P, Lamont RJ et al (2006) Intercellular spreading of *Porphyromonas gingivalis* infection in primary gingival epithelial cells. Infect Immun 74(1):703–710

Yoo C, Ryu MH, Huh J et al (2007) *Chlamydia psittaci* infection and clinicopathologic analysis of ocular adnexal lymphomas in Korea. Am J Hematol 82(9):821–823

Zarkin BA, Lillemoe KD, Cameron JL et al (1990) The triad of *Streptococcus bovis* bacteremia, colonic pathology, and liver disease. Ann Surg 211(6):786–791

Zatloukal P, Kubík A, Pauk N et al (2003) Adenocarcinoma of the lung among women: risk associated with smoking, prior lung disease, diet and menstrual and pregnancy history. Lung Cancer 41(3):283–293

Zhan P, Suo LJ, Qian Q et al (2011) *Chlamydia pneumoniae* infection and lung cancer risk: a meta-analysis. Eur J Cancer 47(5):742–747

Zhang GS, Winter JN, Variakojis D et al (2007) Lack of an association between *Chlamydia psittaci* and ocular adnexal lymphoma. Leuk Lymphoma 48(3):577–583

Zhang DD, Liu HG, Li HY et al (2009) Relationship between primary ocular adnexal mucosa-associated lymphoid tissue lymphoma and eye infection. Zhonghua Bing Li Xue Za Zhi 38(8):513–518

Zhang D, Dong L, Li H et al (2010) Ocular adnexal mucosa-associated lymphoid tissue lymphoma in Northern China: high frequency of numerical chromosomal changes and no evidence of an association with *Chlamydia psittaci*. Leuk Lymphoma 51(11):2031–2038

Zhang LJ, Hong L, Chen N et al (2011) Promoting effect of *Chlamydia pneumoniae* infection on human laryngeal carcinoma HEp-2 cell adhesion and migration. Zhonghua Zhong Liu Za Zhi 33(1):18–22

Zheng W, Blot WJ, Liao ML et al (1987) Lung cancer and prior tuberculosis infection in Shanghai. Br J Cancer 56(4):501–504

Chapter 4
The Role of Protozoa in Cancer Development

Abstract The role of protozoa in cancer development remains obscure, and to the moment of the publication of this book, no one of them were recognized IARC as a causative agent of human cancer. However, available information from original research and reviews reflects a possibility to cause cancer for certain *Protozoa* species. We start from *Toxoplasma gondii* through *Cryptosporidium parvum* to *Trichomonas vaginalis*.

Keywords Cancer • Infections • Infectious agents • Carcinogens • Biological carcinogenesis • Protozoa • Toxoplasma • Cryptosporidium • Trichomonas • Toxins

4.1 Toxoplasma gondii

Toxoplasma gondii (*T. gondii*) is a protozoan, obligate intracellular parasite, with felines as a definitive host. In humans who have their immunity intact, the infection is usually benign, owing to a rapid immune response that forces encystation of the tachyzoites. Nevertheless, viable organisms may remain encysted within the host for extended periods of time. Encysted organisms may serve as reservoirs for relapse of disease in the conditions of compromised cellular immunity.

The first researchers who suspected the possible role of *Toxoplasma gondii* (*T. gondii*) in cancer development were Schuman et al. (1967). In their study (126 U.S. cases, 126 controls), the correlation between antibodies against *T. gondii* and the risk of the central nervous system neoplasms was observed (particularly for gliomas as a whole and for astrocytomas as a subgroup). Rocchi et al. (1975) detected that the incidence of high titer of anti-*T. gondii* antibodies was greater in patients with Hodgkin's disease in comparison with controls. Ortiz-Muñoz et al. (1984) showed the incidence and titer of antibodies to *T. gondii* in the sera of 63 patients with non-Hodgkin lymphoma being significantly higher in non-Hodgkin lymphoma patients. Mueller et al. (1988) evaluated the prevalence of anti-*T. gondii*

A.G. Kutikhin et al., *Infectious Agents and Cancer*, DOI 10.1007/978-94-007-5955-8_4,

antibodies in 50 patients with seminoma, 50 subjects with embryonal carcinoma, 17 with other sub-classifications, and 100 matched controls, but no significant differences was noted. Ryan et al. (1993) collected blood from 117 Australian subjects with glioma, 53 individuals with meningioma and 415 controls, and assessed the presence of IgG antibodies to *T. gondii*, but there were no statistically significant differences found. In the Adelaide region (South Australia) an increased risk of meningioma in *T. gondii* carriers was observed (OR=2.09, 95% CI=1.14–3.83). Huang et al. (2000) investigated the prevalence of anti-*T. gondii* IgG, IgM antibodies and circulating antigens in 50 Chinese cervical cancer cases, 59 patients with gynaecological benign tumors and 138 subjects with other diseases. According to their data, the total positive rates of the circulating antigens and antibodies in all three groups were 44.0% (22 of 50), 25.4% (15 out of 59) and 20.3% (28 of 138), so the *T. gondii*-infection rate of the cervical cancer group was higher than those of the other two groups, particularly for the positive rates of circulating antigens and IgM. Shen et al. (2001) analyzed ten Swiss or U.S. cases of primary intraocular lymphoma using microdissection and PCR and detected *T. gondii* DNA in the lymphoma but not in normal cells of two cases that resembled ocular toxoplasmosis clinically, proposing the possible role of *T. gondii* in the development of some forms of primary intraocular B-cell lymphoma. Rai et al. (2003) reported that Nepal patients with ocular malignancies had the highest positive rate [68.7% (22 of 32 affected subjects)] compared to other subjects with ocular diseases (240 individuals). Yazar et al. (2004) detected anti-*T. gondii* IgG antibodies in 68 out of 108 (63%) Turkish cancer patients and only in 21 of 108 (19.4%) of the controls, that reached the level of statistical significance. In addition, anti-*T. gondii* IgM antibodies were detected in seven (6.5%) patients and in one (0.9%) control. Yuan et al. (2007) revealed higher anti-*T. gondii* IgG antibodies positivity rates in 267 Chinese cancer patients compared to controls. It should be noted that the positivity rates of anti-*T. gondii* IgG in nasopharyngeal carcinoma and rectal cancer were significantly higher than the other cancer groups, but the differences in IgM positivity rates were not significant. Shin et al. (2009) also noted the high titers of IgG antibodies against *T. gondii* in patients with neoplasms in their screening investigation in Korea. Thomas et al. (2012) predicted that *T. gondii* can increase the risk of brain cancer provoking inflammation and inhibiting apoptosis and compared national incidence of brain malignancies and seroprevalence of *T. gondii*. The prevalence of *T. gondii* explained 19% of the residual variance in brain cancer incidence after controlling for the positive effects of gross domestic product and latitude among nations, and infection with *T. gondii* was associated with a 1.8-fold increase in the risk of brain cancers. In the next study of this research group, Thomas et al. (2012) and Vittecoq et al. (2012) detected that in France from the ages of 55 and older, regional mortality rates due to brain cancer correlated positively with the local seroprevalence of *T. gondii*, and this effect was particularly strong for men.

In addition, Nash et al. (1998) reported that cells infected with *T. gondii* are resistant to multiple inducers of apoptosis, including Fas-dependent and Fas-independent CTL-mediated cytotoxicity, IL-2 deprivation, gamma irradiation, UV irradiation, and the calcium ionophore beauvericin. The inhibition of apoptosis required live intracellular parasite and ongoing protein synthesis. The authors suggested that one common mechanism should be involved in the inhibition of such apoptotic pathways.

To sum up, there are extensive epidemiological evidences that *T. gondii* may be a cause of various malignant tumors, but there are almost only retrospective seroepidemiological studies, and it is difficult to assess the cause of this relationship. According to our criteria, it is possible to include it in the extended list of the infectious agents that may be a cause of cancer development but are not placed in the IARC roster.

4.2 Cryptosporidium parvum

Cryptosporidiosis represents a major public health problem. This infection, caused by protozoa of the genus *Cryptosporidium*, has been reported worldwide as a frequent cause of diarrhoea, and its prevalence varies according to different regions. Sulzyc-Bielicka et al. (2007) performed parasitological examination of feces for 55 colorectal cancer patients and did not detect in the specimens parasites other than *Cryptosporidium* spp. Coproantigens of this parasite were detected primarily in the patients with tumors located in the sigmoid and descending colon. Certad et al. (2007) developed a model using adult SCID mice inoculated with *Cryptosporidium parvum* (*C. parvum*) or *Cryptosporidium muris* (*C. muris*) while being treated with or without dexamethasone. Among *C. parvum*-infected mice, dexamethasone-treated SCID mice became chronic shedders with a prepatent period of 6.2 days on average and developed glandular cystic polyps with areas of intraepithelial neoplasia and with the presence of intramucosal adenocarcinoma. In further investigations of Certad et al. (2010a, b), SCID mice infected with *C. parvum* TUM1 strain developed a fulminant cryptosporidiosis associated with intramucosal adenocarcinoma. SCID mice treated with dexamethasone were challenged with (*C. parvum*) oocysts. Using Ki-67 as a marker of cellular proliferation, the authors found that mice receiving higher inocula (10^6–10^7) were characterized by more severe neoplastic development, not only in the caecum but in the stomach and duodenum as well. Interestingly, SCID mice inoculated with 10^5–10^7 oocysts had high grade intraepithelial neoplasia or adenomas with high grade dysplasia in the caecum after Day 46 post-infection (PI). Immunohistochemistry for Ki-67 staining indicated that the neoplastic process is associated with cryptosporidiosis and it showed the first immunohistochemical alterations at early stages of the process. In 2012, Certad et al. (2012) reported a case of a patient with fulminant cryptosporidiosis caused by *C. parvum*, which, being inoculated into SCID mice, induced invasive gastrointestinal and biliary adenocarcinoma. These results indicated for the first time that a human-derived *C. parvum* isolate is able to induce cancer. Finally, Shebl et al. (2012) analyzed 320 colorectal cancer cases occurred among 471,909 people with AIDS. Following cryptosporidiosis, colorectal cancer risk was elevated at borderline significance (HR = 2.08, 95% CI = 0.93–4.70).

To summarize, there are basic as well as epidemiological evidence that *C. parvum* may be a cause of colorectal cancer, but the number of investigations carried out to date is rather small. According to our criteria, it is possible to include it in the extended list of the infectious agents that may be a cause of cancer development but are not placed in the IARC roster.

4.3 Trichomonas vaginalis

Trichomonas vaginalis (*T. vaginalis*), a protozoan parasite, is the causative agent of trichomoniasis, the most common nonviral sexually transmitted disease in humans. This parasite has a worldwide distribution and it infects 250–350 million people worldwide. The question about the association of *T. vaginalis* infection and cervical cancer risk was discussed from the late 1950s and early 1960s. The meta-analysis of Zhang and Begg (1994) based on the results of 24 studies published by Berggren (1969), Thomas (1973), Kolinski (1975), Kauraniemi et al. (1978), Thomas and Rawls (1978), Mekki and Ivić (1979), Fedi et al. (1979), Pillay and Yap (1979), Averna et al. (1980), Franceschi et al. (1983), Kazanowska and Kuczyńska (1985), Guijon et al. (1985), Zarković (1985), La Vecchia (1985), Herman-Giddens et al. (1988), Alberico et al. (1988), Meijer et al. (1989), Slattery et al. (1989), Moscicki et al. (1989), Omer et al. (1991), Gram et al. (1992), Chakrabarti et al. (1992), Guijon et al. (1992), Kharsany et al. (1993) found a RR of 1.93 (95% CI = 1.22–2.65) based on the results of two published cohort studies before 1994. The results of the 22 retrospective studies were much less consistent but most of them demonstrated a significant positive association. After 1994, Yap et al. (1995) collected serum samples and evaluated anti-*T. vaginalis* antibodies in 121 invasive cervical cancer patients from Singapore and 242 random age-matched female controls. Antibodies to *T. vaginalis* were revealed in 41.3% (50 of 121) of invasive cervical cancer patients compared with only 5% (12 out of 242) of controls, so *T. vaginalis* significantly increased cervical cancer risk (RR = 3.42, 95% CI = 1.73–6.78). Thereafter, Zhang et al. (1995) observed a relationship between *T. vaginalis* infection and cervical cancer in the prospective study in a cohort of 16,797 Chinese women. *T. vaginalis*-infection correlated with higher cervical cancer risk (RR = 3.3, 95% CI = 1.5–7.4). In a large cohort study conducted in Finland by Viikki et al. (2000) *T. vaginalis* was associated with a high RR of cervical cancer, 6.4 (95% CI = 3.7–10) and SIR = 5.5 (95% CI = 4.2–7.2s), respectively. Sayed el-Ahl et al. (2002) examined a relationship between *T. vaginalis* and cervical cancer by analyzing the sera of 48 invasive cervical cancer Egyptian patients and 100 matched controls. Antibodies to *T. vaginalis* were detected in sera of 18.75% (nine out of 48) of cervical cancer patients and in 5% (five of 100) of controls, and more than 3-fold increased cervical cancer risk was detected for *T. vaginalis* carriers. A cohort study of Samoff et al. (2005) included 621 U.S. female subjects which did not reveal the correlation between *T. vaginalis* and cervical cancer, but the results could be confounded by relatively insensitive technique. Zhao et al. (2006) examined the association of

Table 4.1 The connection of *T. vaginalis* with prostate cancer in epidemiological studies

Authors, year, population	Sample size	Methods of detection	Share of positive individuals, OR, 95% CI
Sutcliffe et al. (2006), U.S. population	691 cases, 691 controls	ELISA	12.5% (cases), 9.5% (controls) 1.43 (1.00–2.03) IgG is the indicator
Stark et al. (2009), U.S. population	673 cases, 673 controls	ELISA	24.5% (cases), 21.4% (controls) 1.23 (0.94–1.61), non-significant For extraprostatic prostate cancer: 2.17 (1.08–4.37) For metastatic cancer: 2.69 (1.37–5.28) IgG is the indicator
Sutcliffe et al. (2009), U.S. population	616 cases, 616 controls	ELISA	21% (cases), 25% (controls) For men with low seropositivity: 0.83 (0.63–1.09) For men with high positivity: 0.97 (0.70–1.34) IgG is the indicator

risk factors to cervical cancer in 8,798 unscreened Chinese women, revealing that the OR of oncogenic HPV-infection was 0.82 (95% CI=0.72–0.94) for subjects suffering from vaginal trichomoniasis.

Sutcliffe et al. (2006) investigated the relation between *T. vaginalis* and prostate cancer in 691 U.S. cases and 691 controls, and found 13% cases and 9% of controls seropositive for trichomonosis (OR=1.43, 95% CI=1.00–2.03). This association persisted after additional adjustment for such factors as a history of other STIs, and was strongest for men who used aspirin infrequently over the course of their lives (OR=2.05; 95% CI=1.05–4.02). Then, Stark et al. (2009) conducted a case-control study including 673 U.S. prostate cancer patients and 673 matched controls. The connection of *T. vaginalis*-seropositive status and prostate cancer risk (OR=1.23, 95% CI=0.94–1.61) was non-significant. In addition, a seropositive status was associated with statistically significantly increased risk of extraprostatic prostate cancer (OR=2.17, 95% CI=1.08–4.37) and of metastatic cancer or prostate cancer-specific death (OR=2.69, 95% CI=1.37–5.28). Thereafter, Sutcliffe et al. (2009) recruited 616 prostate cancer cases and 616 matched U.S. controls compared to seronegative men, the OR of prostate cancer for men with low seropositivity was 0.83 (95% CI=0.63–1.09), and for men with high seropositivity it was 0.97 (95% CI=0.70–1.34). Groom et al. (2012) collected prostate DNA samples from 100 U.K. patients, but failed to detect the presence of any of these agents in investigated samples (Table 4.1).

In addition, Mekki and Ivić (1979), detected that *T. vaginalis* were of a significantly smaller diameter in invasive carcinoma and carcinoma in situ in comparison with dysplasia. In the control group with trichomoniasis alone, the diameter of *T. vaginalis* was twice as large as that in carcinoma and larger compared to dysplasia, indicating that small forms of *T. vaginalis* are more carcinogenic than large ones. Da Costa et al. (2005) assessed the modifications at the junctional complex level of Caco-2 cells after interaction with *T. vaginalis*. The authors showed that *T. vaginalis* adhered to the epithelial cell causing alterations in the junctional complex (decrease in transepithelial electrical resistance, alteration in the pattern of junctional complex proteins distribution as observed for E-cadherin, occludin and ZO-1, enlargement of the spaces between epithelial cells). These effects were dependent on the degree of the parasite virulence isolate, on the iron concentration in the culture medium, and on the expression of adhesin proteins on the parasite surface. According to the data of Fichorova et al. (2006), *T. vaginalis* purified lipophosphoglycan mediated the adhesion of parasites to human vaginal epithelial cells in a dose-dependent manner and induced a selective upregulation of chemotactic cytokines by human endocervical, ectocervical, and vaginal epithelial cells, which do not express Toll-like receptor 4/MD2. In addition, lipophosphoglycan dose-dependently triggered IL-8, promoting the adhesion and transmigration of neutrophils across the endothelium, and macrophage inflammatory protein 3α, which is a chemoattractant for immune cells and is essential for dendritic cell maturation.

To sum up, there are basic as well as epidemiological evidence that *T. vaginalis* may be a cause of cervical and prostate cancer, but with distinct level of reliability: For cervical cancer it is evident, for prostate cancer it is arguable. According to our criteria, it is possible to include it in the short list of the infectious agents that may be a cause of cancer development but are not placed in the IARC roster.

References

Alberico S, Facca MC, Di Bonito L et al (1988) Frequency of cervico-vaginal infections (Trichomonas vaginalis; Chlamydia trachomatis -CHL-; herpes simplex virus -HSV-; human papilloma virus -HPV-) in cervical intraepithelial neoplasia. Eur J Gynaecol Oncol 9(3):252–257

Averna R, Martelli D, Migliorini D et al (1980) Mycoplasmas and dysplasia of the uterine cervix. Boll Ist Sieroter Milan 59(4):348–358

Berggren O (1969) Association of carcinoma of the uterine cervix and Trichomonas vaginalis infestations. Frequency of trichomonas vaginalis in preinvasive and invasive cervical carcinoma. Am J Obstet Gynecol 105(2):166–168

Certad G, Ngouanesavanh T, Guyot K et al (2007) *Cryptosporidium parvum*, a potential cause of colic adenocarcinoma. Infect Agent Cancer 2:22

Certad G, Creusy C, Guyot K et al (2010a) Fulminant cryptosporidiosis associated with digestive adenocarcinoma in SCID mice infected with *Cryptosporidium parvum* TUM1 strain. Int J Parasitol 40(13):1469–1475

Certad G, Creusy C, Ngouanesavanh T et al (2010b) Development of *Cryptosporidium parvum*-induced gastrointestinal neoplasia in severe combined immunodeficiency (SCID) mice: severity of lesions is correlated with infection intensity. Am J Trop Med Hyg 82(2):257–265

Certad G, Benamrouz S, Guyot K et al (2012) Fulminant cryptosporidiosis after near-drowning: a human *Cryptosporidium parvum* strain implicated in invasive gastrointestinal adenocarcinoma and cholangiocarcinoma in an experimental model. Appl Environ Microbiol 78(6):1746–1751

Chakrabarti RN, Dutta K, Sarkhel T et al (1992) Cytologic evidence of the association of different infective lesions with dysplastic changes in the uterine cervix. Eur J Gynaecol Oncol 13(5):398–402

da Costa RF, de Souza W, Benchimol M et al (2005) Trichomonas vaginalis perturbs the junctional complex in epithelial cells. Cell Res 15(9):704–716

Fedi B, Marchetti E, Pelini G et al (1979) Association between Trichomonas vaginalis and some types of lesions of the uterine neck based on data collected in the quinquennium of 1974–1978. Boll Soc Ital Biol Sper 55(13):1300–1303

Fichorova RN, Trifonova RT, Gilbert RO et al (2006) Trichomonas vaginalis lipophosphoglycan triggers a selective upregulation of cytokines by human female reproductive tract epithelial cells. Infect Immun 74(10):5773–5779

Franceschi S, Doll R, Gallwey J et al (1983) Genital warts and cervical neoplasia: an epidemiological study. Br J Cancer 48(5):621–628

Gram IT, Macaluso M, Churchill J et al (1992) Trichomonas vaginalis (TV) and human papillomavirus (HPV) infection and the incidence of cervical intraepithelial neoplasia (CIN) grade III. Cancer Causes Control 3(3):231–236

Groom HC, Warren AY, Neal DE et al (2012) No evidence for infection of UK prostate cancer patients with XMRV, BK virus, Trichomonas vaginalis or human papilloma viruses. PLoS One 7(3):e34221

Guijon FB, Paraskevas M, Brunham R (1985) The association of sexually transmitted diseases with cervical intraepithelial neoplasia: a case-control study. Am J Obstet Gynecol 151(2):185–190

Guijon FB, Paraskevas M, Rand F et al (1992) Vaginal microbial flora as a cofactor in the pathogenesis of uterine cervical intraepithelial neoplasia. Int J Gynaecol Obstet 37(3):185–191

Herman-Giddens ME, Gutman LT, Berson NL (1988) Association of coexisting vaginal infections and multiple abusers in female children with genital warts. Sex Transm Dis 15(1):63–67

Huang H, Yan FH, Li CJ et al (2000) Detection of Toxoplasma infection in women with gynaecologic neoplasms using ELISA. Zhongguo Ji Sheng Chong Xue Yu Ji Sheng Chong Bing Za Zhi 18(3):165–166

Kauraniemi T, Räsänen-Virtanen U, Hakama M (1978) Risk of cervical cancer among an electrocoagulated population. Am J Obstet Gynecol 131(5):533–538

Kazanowska W, Kuczyńska K (1985) Occurrence of Trichomonas vaginalis infection in patients with neoplasms of the female genitalia. Wiad Parazytol 31(4–6):567–570

Kharsany AB, Hoosen AA, Moodley J et al (1993) The association between sexually transmitted pathogens and cervical intra-epithelial neoplasia in a developing community. Genitourin Med 69(5):357–360

Kolinski J (1975) Correlation between trichomoniasis and cervix neoplasms in women. Ginekol Pol 46(2):181–186

La Vecchia C (1985) The epidemiology of cervical neoplasia. Biomed Pharmacother 39(8):426–433

Meijer CJ, Calame JJ, de Windt EJ et al (1989) Prevalence of Chlamydia trachomatis infection in a population of asymptomatic women in a screening program for cervical cancer. Eur J Clin Microbiol Infect Dis 8(2):127–130

Mekki F, Ivić J (1979) Significance of variations in the size of Trichomonas vaginalis in patients with dysplasia, intrapithelial and invasive planocellular carcinoma of the uterine cervix. Jugosl Ginekol Opstet 18(1):15–19

Moscicki AB, Winkler B, Irwin CE Jr et al (1989) Differences in biologic maturation, sexual behavior, and sexually transmitted disease between adolescents with and without cervical intraepithelial neoplasia. J Pediatr 115(3):487–493

Mueller N, Hinkula J, Wahren B (1988) Elevated antibody titers against cytomegalovirus among patients with testicular cancer. Int J Cancer 41(3):399–403

Nash PB, Purner MB, Leon RP et al (1998) Toxoplasma gondii-infected cells are resistant to multiple inducers of apoptosis. J Immunol 160(4):1824–1830

Omer EE, Naeem HA, Ali MH et al (1991) Vaginal and cervical abnormalities associated with trichomonal infection. East Afr Med J 68(6):455–460

Ortiz-Muñoz AB, Cuadrado-Méndez L, Sanchis-Belenguer R (1984) Possible interactions between Toxoplasma gondii infection and the presence of non-Hodgkin's lymphoma. Rev Esp Oncol 31(2):237–245

Pillay B, Yap SK (1979) Trichomoniasis – incidence in pill users and associated Pap smear abnormalities. Malays J Pathol 2:59–62

Rai SK, Upadhyay MP, Shrestha HG (2003) Toxoplasma infection in selected patients in Kathmandu, Nepal. Nepal Med Coll J 5(2):89–91

Rocchi G, Tosato G, Papa G (1975) Antibodies to Epstein-Barr virus-associated nuclear antigen and to other viral and non-viral antigens in Hodgkin's disease. Int J Cancer 16(2):323–328

Ryan P, Hurley SF, Johnson AM et al (1993) Tumours of the brain and presence of antibodies to Toxoplasma gondii. Int J Epidemiol 22(3):412–419

Samoff E, Koumans EH, Markowitz LE et al (2005) Association of Chlamydia trachomatis with persistence of high-risk types of human papillomavirus in a cohort of female adolescents. Am J Epidemiol 162(7):668–675

Sayed el-Ahl SA, el-Wakil HS, Kamel NM et al (2002) A preliminary study on the relationship between Trichomonas vaginalis and cervical cancer in Egyptian women. J Egypt Soc Parasitol 32(1):167–178

Schuman LM, Choi NW, Gullen WH (1967) Relationship of central nervous system neoplasms to Toxoplasma gondii infection. Am J Public Health Nations Health 57(5):848–856

Shebl FM, Engels EA, Goedert JJ (2012) Opportunistic intestinal infections and risk of colorectal cancer among people with AIDS. AIDS Res Hum Retroviruses 28(9):794–799

Shen DF, Herbort CP, Tuaillon N et al (2001) Detection of Toxoplasma gondii DNA in primary intraocular B-cell lymphoma. Mod Pathol 14(10):995–999

Shin DW, Cha DY, Hua QJ et al (2009) Seroprevalence of Toxoplasma gondii infection and characteristics of seropositive patients in general hospitals in Daejeon, Korea. Korean J Parasitol 47(2):125–130

Slattery ML, Overall JC Jr, Abbott TM et al (1989) Sexual activity, contraception, genital infections, and cervical cancer: support for a sexually transmitted disease hypothesis. Am J Epidemiol 130(2):248–258

Stark JR, Judson G, Alderete JF et al (2009) Prospective study of Trichomonas vaginalis infection and prostate cancer incidence and mortality: Physicians' Health Study. J Natl Cancer Inst 101(20):1406–1411

Sulzyc-Bielicka V, Kuźna-Grygiel W, Kołodziejczyk L et al (2007) Cryptosporidiosis in patients with colorectal cancer. J Parasitol 93(3):722–724

Sutcliffe S, Giovannucci E, Alderete JF et al (2006) Plasma antibodies against Trichomonas vaginalis and subsequent risk of prostate cancer. Cancer Epidemiol Biomarkers Prev 15(5):939–945

Sutcliffe S, Alderete JF, Till C et al (2009) Trichomonosis and subsequent risk of prostate cancer in the Prostate Cancer Prevention Trial. Int J Cancer 124(9):2082–2087

Thomas DB (1973) An epidemiologic study of carcinoma in situ and squamous dysplasia of the uterine cervix. Am J Epidemiol 98(1):10–28

Thomas DB, Rawls WE (1978) Relationship of herpes simplex virus type-2 antibodies and squamous dysplasia to cervical carcinoma in situ. Cancer 42(6):2716–2725

Thomas F, Lafferty KD, Brodeur J et al (2012) Incidence of adult brain cancers is higher in countries where the protozoan parasite Toxoplasma gondii is common. Biol Lett 8(1):101–103

Viikki M, Pukkala E, Nieminen P et al (2000) Gynaecological infections as risk determinants of subsequent cervical neoplasia. Acta Oncol 39(1):71–75

Vittecoq M, Elguero E, Lafferty KD et al (2012) Brain cancer mortality rates increase with Toxoplasma gondii seroprevalence in France. Infect Genet Evol 12(2):496–498

Yap EH, Ho TH, Chan YC et al (1995) Serum antibodies to Trichomonas vaginalis in invasive cervical cancer patients. Genitourin Med 71(6):402–404

Yazar S, Yaman O, Eser B et al (2004) Investigation of anti-Toxoplasma gondii antibodies in patients with neoplasia. J Med Microbiol 53(Pt 12):1183–1186

Yuan Z, Gao S, Liu Q et al (2007) Toxoplasma gondii antibodies in cancer patients. Cancer Lett 254(1):71–74

Zarković G (1985) Alterations of cervical cytology and steroid contraceptive use. Int J Epidemiol 14(3):369–377

Zhang ZF, Begg CB (1994) Is Trichomonas vaginalis a cause of cervical neoplasia? Results from a combined analysis of 24 studies. Int J Epidemiol 23(4):682–690

Zhang ZF, Graham S, Yu SZ et al (1995) Trichomonas vaginalis and cervical cancer. A prospective study in China. Ann Epidemiol 5(4):325–332

Zhao FH, Forman MR, Belinson J et al (2006) Risk factors for HPV infection and cervical cancer among unscreened women in a high-risk rural area of China. Int J Cancer 118(2):442–448

Chapter 5
The Role of Helminthes and Fungi in Cancer Development

Abstract At the moment of publication, IARC recognized *Schistosoma haematobium et mansoni*, *Opisthorchis viverrini*, and *Clonorchis sinensis* as causative agents of cancer, leaving a possibility to enlarge this list by Schistosoma japonicum et Opisthorchis felineus. However, there are certain additional helminthes which can cause human cancer, and we describe the available relevant information on them in this chapter. In addition, fungi may be also associated with cancer, although the main part of information about them as a cancer cause will be given in the next chapter due to the concept of this book.

Keywords Cancer • Infections • Infectious agents • Carcinogens • Biological carcinogenesis • Helminthes • Fungi • Taenia • Strongyloides • Fasciola

5.1 Taenia spp.

In the study of Hanes and Stribling (1995) two of five male Sprague-Dawley rats with hepatic tapeworm cysts developed large multinodular fibrosarcomas, and the parasites were identified as *Cysticercus fasciolaris*, the larval stage of *Taenia taeniaeformis*. The development of sarcomas in rats induced by *Taenia sp*. was caused by chronic inflammatory reaction of the capsule. Del Brutto et al. (1997) recruited 43 patients with a cerebral glioma and 172 matched controls, investigating the association between neurocysticercosis and this malignancy. In this study, the OR was 7.63 (95% CI = 2.03–31.09) thus suggesting neurocysticercosis as a risk factor for cerebral glioma. Herrera et al. (1999) evaluated 1,271 autopsy files, revealing that the frequency of neurocysticercosis was higher in patients with malignant hematological diseases compared to controls. The OR for this association was 3.54 (95% CI = 1.17–9.79). In the following year this research group [Herrera et al. (2000)] investigated whether DNA damage arises in peripheral lymphocytes in

A.G. Kutikhin et al., *Infectious Agents and Cancer*, DOI 10.1007/978-94-007-5955-8_5,

patients with neurocysticercosis and showed that the frequencies of chromosome aberrations and micronuclei in peripheral blood lymphocytes were higher in the infected patients as compared to healthy donors. Del Brutto et al. (2000) proposed the following pathogenetic mechanisms explaining possible oncogenic effects of cysticerci: Parasite-induced modulation of the host immune response that may be associated with loss of regulatory mechanisms involved in the immunological surveillance against cancer; transfer of genetic material from the parasite to the host, causing DNA damage and malignant transformation of host cells; chronic inflammation with liberation of nitric oxide and inhibition of tumor suppressor genes. Herrera et al. (2003) discovered a protein secreted by the metacestode of *T. solium* capable of transforming the Syrian hamster embryo cells. In addition, the frequency of micronucleated cells in lymphocytes treated with MF was significantly increased. Silva et al. (2006) demonstrated the genotoxicity of the vesicular fluid and saline extract of *T. solium*.

In summary, there are basic as well as epidemiological evidence that *T. solium* may be a cause of glioma or hematological malignancies, but this is arguable. According to our criteria, it is possible to include it in the extended list of the infectious agents that may be a cause of cancer development but are not placed in the IARC roster.

5.2 Strongyloides stercoralis

Strongyloides stercoralis (*S. stercoralis*) causes strongyloidiasis, a chronic, usually asymptomatic, gastrointestinal infection. In the study of Gabet et al. (2000), the mean circulating HTLV-1 proviral load was more than 5-fold higher in HTLV-1 carriers with strongyloidiasis than in HTLV-1+ individuals without *S. stercoralis*-infection ($P<0.009$). This increased proviral load was found to result from oligoclonal expansion of infected cells, but such positive effect of *S. stercoralis* on clonal expansion was reversible under effective treatment of strongyloidiasis. Hirata et al. (2007) investigated the association of *S. stercoralis* infection and the hepatopancreatobiliary cancer risk by recruiting 196 Japanese patients with hepato-pancreato-biliary cancer and 1,458 matched controls. It was revealed by the authors that *S. stercoralis*-infection correlated with increased cancer risk (OR = 2.7, 95% CI = 1.1–6.3). Lin et al. (2008) examined 943 Kaposi sarcoma patients and 1,042 other cancer patients and observed that patients with Kaposi sarcoma were characterized by higher carriage of *S. stercoralis* larvae (OR = 2.1, 95% CI = 1.2–3.7).

To summarize, there are basic as well as epidemiological evidence that *T. solium* may be a cause of glioma or hematological malignancies, but this is arguable. According to our criteria, it is possible to include it in the extended list of the infectious agents that may be a cause of cancer development but are not placed in the IARC roster.

5.3 Fasciola hepatica, Trichostrongylus colubriformis, Paracoccidioides brasiliensis

Tsocheva et al. (1992) assessed the activity of the rat liver monooxygenase system after a single and combined treatment with *Fasciola hepatica* and diethylnitrosamine, found the changes in drug metabolism was being caused mainly by *F. hepatica*. Montero et al. (1999) demonstrated an elevated activity of the CYP2A5 isozyme, increased inflammation and proliferation in male mouse liver infected with *F. hepatica*. Motorna et al. (2001) investigated whether *F. hepatica*-induced inflammation could induce mutagenic events in mammalian tissue. An increase in lacI mutations in liver cells harvested from mice harboring *F. hepatica* flukes was noted when they compared them to uninfected control animals. Complex changes and multiple mutations were elevated in the spectrum of mutations from *F. hepatica*-infected animals (18.2%) in comparison with control animals (2.8%). In the investigation of Huby et al. (1995) excretory/secretory (ES) products of the nematode parasite *Trichostrongylus colubriformis* caused an increase in vitro proliferation of the epithelial cell line HT29-D4. In the next study of this research group [Huby et al. (1999)], the excretory/secretory products of *T. colubriformis* increased the proliferation of three epithelial intestinal cell lines (RIC, IEC-6, IRD-98) and of epithelial kidney cell line (MDCK) but inhibited the proliferation of epithelial ovarian cell line (CHO) and fibroblasts (3T3). Rodrigues Gda et al. (2010) reviewed 25 cases diagnosed with paracoccidiomycosis and cancer and detected that this condition, caused by *Paracoccidioides brasiliensis*, may increase lung cancer risk.

In summary, there are few basic evidences that *F. hepatica*, *T. colubriformis*, and *Paracoccidioides brasiliensis* may be a cause of cancer. According to our criteria, it is possible to include them in the extended list of the infectious agents that may be a cause of cancer development but are not placed in the IARC roster.

References

Del Brutto OH, Castillo PR, Mena IX et al (1997) Neurocysticercosis among patients with cerebral gliomas. Arch Neurol 54(9):1125–1128

Del Brutto OH, Dolezal M, Castillo PR et al (2000) Neurocysticercosis and oncogenesis. Arch Med Res 31(2):151–155

Gabet AS, Mortreux F, Talarmin A et al (2000) High circulating proviral load with oligoclonal expansion of HTLV-1 bearing T cells in HTLV-1 carriers with strongyloidiasis. Oncogene 19(43):4954–4960

Hanes MA, Stribling LJ (1995) Fibrosarcomas in two rats arising from hepatic cysts of *Cysticercus fasciolaris*. Vet Pathol 32(4):441–444

Herrera LA, Benita-Bordes A, Sotelo J et al (1999) Possible relationship between neurocysticercosis and hematological malignancies. Arch Med Res 30(2):154–158

Herrera LA, Ramirez T, Rodríguez U et al (2000) Possible association between *Taenia solium* cysticercosis and cancer: increased frequency of DNA damage in peripheral lymphocytes from neurocysticercosis patients. Trans R Soc Trop Med Hyg 94(1):61–65

Herrera LA, Tato P, Molinari JL et al (2003) Induction of DNA damage in human lymphocytes treated with a soluble factor secreted by *Taenia solium* metacestodes. Teratog Carcinog Mutagen 23 (Suppl. 1) 23:79–83

Hirata T, Kishimoto K, Kinjo N et al (2007) Association between *Strongyloides stercoralis* infection and biliary tract cancer. Parasitol Res 101(5):1345–1348

Huby F, Hoste H, Mallet S et al (1995) Effects of the excretory/secretory products of six nematode species, parasites of the digestive tract, on the proliferation of HT29-D4 and HGT-1 cell lines. Epithelial Cell Biol 4(4):156–162

Huby F, Nano JL, Mallet S et al (1999) Effects of the excretory/secretory products of *Trichostrongylus colubriformis* on the growth of different cell lines. Int J Parasitol 29(5):697–702

Lin CJ, Katongole-Mbidde E, Byekwaso T et al (2008) Intestinal parasites in Kaposi sarcoma patients in Uganda: indication of shared risk factors or etiologic association. Am J Trop Med Hyg 78(3):409–412

Montero R, Gentile GJ, Frederick L et al (1999) Induced expression of CYP2A5 in inflamed trematode-infested mouse liver. Mutagenesis 14(2):217–220

Motorna OO, Martin H, Gentile GJ et al (2001) Analysis of lacI mutations in Big Blue transgenic mice subjected to parasite-induced inflammation. Mutat Res 484(1–2):69–76

Rodrigues Gda S, Severo CB, Oliveira Fde M et al (2010) Association between paracoccidioidomycosis and cancer. J Bras Pneumol 36(3):356–362

Silva LP, Costa-Cruz JM, Spanó MA et al (2006) Genotoxicity of vesicular fluid and saline extract of *Taenia solium* metacestodes in somatic cells of *Drosophila melanogaster*. Environ Mol Mutagen 47(4):247–253

Tsocheva NT, Kadiiska MB, Poljakova-Krusteva OT et al (1992) Combined effect of fascioliasis and diethylnitrosamine carcinogenesis on the activity of the rat liver monooxygenase system. Comp Biochem Physiol C 101(3):475–479

Chapter 6
Organ Microbiota in Cancer Development: The Holy Grail of Biological Carcinogenesis

Abstract The problem of the role of human microbiota in cancer development is one of the most attractive in modern biology. It has been proved that microbiota is a key player in a number of vital processes including metabolism, immunity, regulation of apoptosis, growth, proliferation, survival, angiogenesis etc. Our understanding of the role of human microbiota in etiopathogenesis of various diseases has been greatly improved in the recent years, and now microbiota is obviously established as a crucially important part of the human body. Moreover, microbiota possesses both local and distant effects, regulating the homeostasis of various organ systems. However, our understanding of the role of microbiota is only superficial, and a wide spectrum of investigations in this field is required. In this chapter, we shed light on a problem of the significance of oral and gut microbiota (since microbiota of these organs is the most investigated at the moment) in cancer development.

Keywords Cancer • Infections • Infectious agents • Carcinogens • Biological carcinogenesis • Microbiota • Microflora • Oral microbiota • Gut microbiota • Colorectal cancer

6.1 The Role of Oral Microbiota in Cancer Development

Poor oral hygiene may be an independent risk factor for oral cancer, and certain species of oral microbiota were suspected as possible etiological agents of oral vancer. When strains of *Candida* spp. were isolated from leukoplakia lesions and normal mucosa and then evaluated, those with a high potential of producing carcinogenic N-nitroso-benzylmethylamine from salivary precursor molecules were more frequently isolated from lesions in comparison with normal mucosal tissues (Krogh 1990). The polymicrobial burden caused by oral biofilms can be a source of mutagenic and, hence, carcinogenic interactions with saliva (Bloching et al. 2007). Several oral microorganisms (*Candida*spp., *S. salivarius*, *S. intermedius*, *S. mitis*)

A.G. Kutikhin et al., *Infectious Agents and Cancer*, DOI 10.1007/978-94-007-5955-8_6,

are capable of metabolizing highly carcinogenic acetaldehyde from alcohol by alcoholdehydrogenase (ADH) (Tillonen et al. 1999; Homann et al. 2000; Kurkivuori et al. 2007; Nieminen et al. 2009; Uittamo et al. 2009). The carcinogenicity of *C. albicans* was shown in animal models as well by O'Grady and Reade (1992). According to the data of Yumoto et al. (2001), *Eikenella corrodens* is able to stimulate human oral epithelial cells in producing various mediators including IL-6, IL-8, and PGE2, possibly, via the secretion of soluble proteins. *S. anginosus* and *S. mitis* were observed to induce the production of inflammatory cytokines in human esophageal epithelial cell lines (Sasaki et al. 2001; Narikiyo et al. 2004). Sugano et al. (2003) detected the significant elevation in carcinogenic 8-hydroxy-deoxyguanosine (8-OHdG, resulting from the continuous generation of reactive oxygen species associated with chronic inflammation) in patients who tested positive for *S. anginosus* compared to individuals who tested negative for the bacterium. *Mycoplasma (M.) fermentans* and *M. penetrans* have also been shown to prevent apoptosis in vitro (Feng et al. 1999; Gerlic et al. 2004).

In addition, there were a number of fundamental studies analyzing carcinogenic properties of *Porphyromonas (P.) gingivalis*. This bacterium can attach itself to various gingival carcinoma cell lines (Hirose et al. 1996), invade gingival epithelial cells in vitro (Kim et al. 2010), prevent and inhibit apoptosis of epithelial cells (Nakhjiri et al. 2001; Mao et al. 2007), multiply, survive into epithelial cells and spread into other cells (Yilmaz et al. 2006), promote cell survival and increase cell proliferation (Kuboniwa et al. 2008). *P. gingivalis* may also induce COX-2 expression (Kuramitsu et al. 2002) and increase production of TNF-α, IL-6, IL-8, and IL-1b (Andrian et al. 2004). It was also shown that *P. gingivalis* contains proteins and lipopolysaccharides on its outer surface that can stimulate human fibroblasts to proliferate (Mihara et al. 1993; Takemura et al. 1998; Putnins et al. 2002). Based on the results of these studies, Katz et al. (2011) declared a hypothesis that *P. gingivalis* may cause oral cancer and revealed that *P. gingivalis* induces neoplastic transformation of the epithelium and is present in the oral squamous cell carcinoma tissues in higher concentration than in normal tissues.

Periodontitis was associated with elevated head and neck cancer risk in the study of Tezal et al. (2009) who recruited a study sample of 266 U.S. cases and 207 controls (OR=4.36, 95% CI=6.01–93.16). In addition, patients with periodontitis had poorly differentiated oral squamous cell carcinoma (OSCC) more frequently than subjects without periodontitis (32.8% and 11.5%, respectively). Candidal leukoplakias may develop into carcinomas in 9–40% of cases (Hooper et al. 2009), and Candida-infected leukoplakias are characterized by a higher rate of malignant transformation in comparison with other types (Reibel 2003). Chronic hyperplastic candidosis develops into a neoplasm in up to 10% of cases (Bartie et al. 2004). Many patients suffering from the autoimmune polyendocrinopathy-candidiasis-ectodermal dystrophy (APECED) have chronic oral candidosis and are prone to oral carcinoma development at a young age (Rautemaa et al. 2007). Narikiyo et al. (2004) examined the bacterial diversity in esophageal carcinoma tissues and in the saliva of healthy people. This group reported the preferential and frequent infection of the *T. denticola*, *S. mitis*, and *S. anginosus* in esophageal cancer patients from different regions of the world (for *T. denticola*, the frequency was 38% in Japanese sample and 46% in sample consisted of Chinese, French, and Italian patients). Their data suggested that these three bacteria could have significant roles in the carcinogenic process of many cases of esophageal cancer by causing inflammation and by promoting the carcinogenic process, and that eradication of these three bacteria may decrease the risk of recurrence. The first evidence of *C. albicans* association with oral cancer in epidemiological studies was demonstrated in 1969 (Cawson 1969). Nagy et al. (1998) reported that *Veillonella* spp., *Fusobacterium* spp., *Prevotella* spp., *Porphyromonas* spp., *Actinomyces* spp., *Clostridium* spp., *Haemophilus* spp., *Enterobacteriaceae* spp. and *Streptococcus* spp. are detected at increased counts at tumor sites. *C. albicans* was revealed at eight of the 21 tumor sites, but at none of control sites. Case reports suggesting the malignant potential of *C. albicans* were also published (Williams et al. 2001). Lu et al. (2002) investigated 42 patients with OSCC before and after operation, the number of bacterial colony forming units found was significantly higher at the OSCC surface than at the healthy oral mucosa. McCullough et al. (2002) recruited 223 patients who underwent biopsies of oral mucosal lesions and showed that the frequency of oral yeast carriage was significantly greater in patients with oral epithelial dysplasia or OSCC compared to others. In addition, patients with oral epithelial dysplasia or OSCC had a higher number of yeast in their oral cavity than patients without any evidence of epithelial dysplasia or neoplasia. The degree of epithelial dysplasia in these patients correlated with higher counts of yeasts in the oral cavity. Morita et al. (2003) detected *S. anginosus* DNA in five of 38 (13%) tissue samples from patients with oral cancer, and Sasaki et al. (2005) obtained tissue specimens of 46 oral cancer patients and three precancerous leukoplakia subjects, detecting *S. anginosus* DNA in OSCC (19 out of 42, 45%) but not in other types of cancer (lymphoma and rhabdomyosarcoma) or leukoplakia samples. *S. anginosus* was solely detected in dental plaque but not in saliva from all 19 *S. anginosus*-positive squamous cell carcinoma cases. The authors concluded that dental plaque can be a reservoir of the *S. anginosus*. Mager et al. (2005) analyzed 45 cases of OSCC and 229 healthy controls, revealing that *Capnocytophaga gingivalis*, *Prevotella melaninogenica* and *Streptococcus mitis* were elevated in the saliva of individuals with oral squamous cell carcinoma compared to controls. The authors further analyzed whether these species can be used as diagnostic markers for oral cancer, revealing 80% sensitivity and 82% specificity of this test. Hooper et al. (2007) identified *Exiguobacterium oxidotolerans*, *Prevotella melaninogenica*, *Staphylococcus aureus*, *Veillonella parvula* and *Micrococcus* spp. within OSCC but not from the control sites. Byakodi et al. (2011) assessed the microbial flora associated with the OSCC surfaces compared to contralateral healthy mucosa and control healthy mucosa, and evaluated the microbiota from the saliva culture in 30 subjects with OSCC and 30 healthy controls. According to their data, OSCC sites harbor significantly more microbial flora (bacteria and yeasts) compared to healthy mucosa, and the median number of colony forming units in tumor tissue was also significantly higher than that of the healthy mucosa, and the similar results were obtained when the saliva of cases and controls were compared. Canković and Bokor-Bratić (2010) showed that prevalence of *Candida* spp. was significantly higher in 30 oral cancer patients (nine out of 30, 30%) compared to 30 control subjects with benign oral mucosal lesions (two of 30, 6.7%). Farrell et al. (2012) evaluated salivary microbiota of 28 patients with pancreatic cancer, 27 subjects suffering from chronic pancreatitis, and 28 matched healthy controls. According to results of their study, 31 bacterial species/clusters were increased in the saliva of patients with pancreatic cancer in comparison to those of the healthy controls whilst 25 bacterial species/clusters were decreased. Two out of six bacterial candidates (*Neisseria elongata* and *S. mitis*) were validated using the independent samples, showing significant variation between pancreatic cancer cases and controls (n=56). Two more bacterial species (*Granulicatella adiacens* and *S. mitis*) showed significant variation between chronic pancreatitis samples and controls. Moreover, the combination of *N. elongata* and *S. mitis* had 96.4% sensitivity and 82.1% specificity in distinguishing patients with pancreatic cancer from healthy subjects.

To sum up, the role of oral microbiota in cancer development is far from the clear determination. In addition, it is also possible that variations on oral microflora may influence risk of other cancer types. There is the necessity to clearly define feasible fundamental carcinogenic mechanisms and propose a list of possible oral biological carcinogens. According to published basic research and epidemiological studies, there is the possibility to distinguish some oral microflora members as possible etiological agents of cancer (*Candida* spp., *S. anginosus*, *S. mitis*, *Prevotella melanogenica*, *Porphyromonas gingivalis*) but a number of further investigations is still needed. All of them may be included in the extended list of the infectious agents that may be a cause of cancer development but are not placed in the IARC roster.

6.2 The Role of Gut Microbiota in Colorectal Cancer Development

According to the recent data (Gill et al. 2006), human gut is colonized by approximately 100 trillions of bacteria belonging to 36,000 species (10^{11} bacteria per gram of feces) consistituting its microbiota. It is considered that number of genes of gut microflora members more than 150-fold exceeds the number of human genes (Qin et al. 2010), that, in combination with the discovery of vital functions of the gut microbiota, suggested *Homo sapiens* as a "superorganism", co-evolutionary developing in parallel to its microflora (Ley et al. 2008). The genome of such "superorganism" can be presented as "metagenome" including *Homo sapiens* genes as well as microbiota genes (Ley et al. 2008). The main challenge to the clear determination of gut microflora composition until recent years was the fact that 70–80% of its members cannot be cultivated in laboratory conditions (Tlaskalová-Hogenová et al. 2011). At once, the modern discoveries in methods of molecular biology provided a possibility to intensively decipher the qualitative and quantitative constitution of the gut microbiota, and stimulated the detailed investigation of the interaction of microbiota members with the host and between each other (Tlaskalová-Hogenová et al. 2011). Basic comparative research led to the understading of the uniqueness of individual microflora, which composition is largely determined during the first years of life but can be altered, reversibly or irreversibly, by many factors (age, environment, diet, drugs, diseases and pathological conditions etc.) (Zoetendal et al. 2008). Bacteria represent the vast majority of the gut microbiota members, and more than 99% of them can be classified into seven main phyla: *Firmicutes*, *Bacteroidetes* (dominating phyla), *Proteobacteria*, *Actinobacteria*, *Fusobacteria*, *Verrucomicrobia*, and *Cyanobacteria* (so-called subdominant phyla) (Bäckhed et al. 2005). These micro organisms play a key role in the maintenance of the local and systemic host homeostasis, producing a number of vitamins and aminoacids, fermenting undigested substrates and endogenicmucus (it leads to the formation of short-chain fatty acids being energy source and stimulating salt and blood absorption), participating in the bile acids biotransformation (that if substantial for cholesterol and glucose metabolism), promoting gut anatomical development, contributing to the immune system development and regulation (particularly, local immune system), preventing gut colonization by various pathogens, stimulating mucus and antimicrobial peptides secretion, consolidating the barrier between the lumen and colorectal tissues, regulating gut vital processes (proliferation, growth, differentiation, survival and death), controlling the angiogenesis in the surrounding tissues and performing a number of other, less important functions (Prakash et al. 2011). Thereby, alterations of the gut microbiota functioning may increase risk of various diseases (infectious, metabolic, autoimmune, oncological, neurological, and endocrine) (Tlaskalová-Hogenová et al. 2011). In this section, the role of gut microbiota in colorectal cancer development will be analyzed. This malignancy is on the third place in the incidence and on the fourth place in mortality in the world, underlining the topicality of the problem (Boyle and Ferlay 2005). In the case with late diagnostics 5-year-survival does not exceed 10%, but early diagnostics may increase it to 90% (Boyle and Ferlay 2005). Hence, it is prospective to develop and implement into the routine medical practice, the new methods of colorectal cancer risk determination for early diagnosis and prevention. As was written above, the new predictive markers of colorectal cancer risk may be found among the alterations of qualitative and quantitative composition of the gut microbiota (Zhang et al. 2011).

The action of metabolic carcinogenic mechanisms is based on the conception that gut microflora members participate in the degradation, modification, biotransformation and detoxificationof chemical compounds coming into the colon after food consumption and its processing in the proximal gastrointestinal tract. For instance, O'Keefe et al. (2009) revealed that colorectal cancer is 65-fold less frequent among African indigenous population in comparison with Afro-Americans. In their investigation, these authors (O'Keefe et al. 2009) found the regional colorectal cancer incidence being dependent, besides other factors, on abundant meat and animal fats consumption leading to increasing of the 7α-dehydroxylating bacteria in the gut of Afro-Americans (particularly, *Clostridium* spp.). It is known that this microorganisms can transform primary bile acids (cholic and xenodeoxycholic) to secondary bile acids (deoxycholic and lytocholic) possessing cytotoxic, mutagenic, and hence carcinogenic action (Bernstein et al. 2005). Sulfation is one of the ways of bile acids detoxification and their enhanced elimination since after this modification they become more water-solubleand less prone to enterohepatic recirculation (O'Keefe 2008). Roessner et al. showed that certain members of *Clostridium* spp., *Bacteroides* spp. and *Bifidobacterium* spp. may form arylsulfatase desulfating transformed bile acids and, thus, promoting carcinogenesis (McBain and Macfarlane 1998). In addition, meat is characterized by large amounts of sulfur-containing amino acids and is a rich source of dietary sulfur stimulating the growth of sulfur-reducing bacteria (*Desulfovibrio* spp., *Desulfobacter* spp., *Desulfobulbus* spp., *Desulfomonas* spp., *Desulfotomaculum* spp.), forming hydrogen sulfide as a result of their activity (O'Keefe 2008). This chemical compound promotes the formation of free radicals, suppresses the utilization of anticarcinogenic butyrate, inhibits mucus secretion and DNA methylation, and, in addition, disruption of its detoxification may also result in colorectal cancer development (Ramasamy et al. 2006).

McBain and Macfarlane (1998) demonstrated that gut-colomizing bacteria can produce glycoside hydrolases (glucosidases) in large amounts leading to formation of aglycons, some of which possess cytotoxic, teratogenic and radiomimetic activity. A number of food chemical compounds are metabolized in the liver, conjugates with glucuronic acid and comes into the colon where it can be transformed to carcinogens by β-glucuronidase-producing bacteria (certain members of *Clostridium* spp.) (Humblot et al. 2007). In the feces of colorectal cancer patients the activity of this enzyme is higher compared to the control group (Kim and Jin 2001). As in the case with β-glucuronidase, azoreductase, produced by many bacteria (particularly, *Clostridium clostridioforme*, *Enterococcus faecalis*, *Ruminococcus obeum*, *Bifidobacterium adolescentis*), metabolizes water-soluble azodyes and some chemotherapeutics to carcinogenic aromatic amines (Azcárate-Peril et al. 2011). Moreover, aromaticamines may also be formed of aromatic nitrocompounds by the bacterial nitroreductases producing in the gut (Roldán et al. 2008). Alcohol dehydrogenase, transforming ethanol to carcinogenic acetaldehyde, can also be produced by gut microbiota members since the acetaldehyde level in the colonic mucosa of gnotobionts was significantly lower in comparison with conventional laboratory rodents (Seitz et al. 1990), and nucleotide sequence encoding this enzyme was found in the majority of known genomes of microorganisms colonizing the gastrointestinal tract (Markowitz et al. 2008). Some *Bacteroidetes* may form fecapentaens, disrupting the antioxidant defense by the gluthatione oxidation and causing the mutations (Huycke and Gaskins 2004). In addition, one of the basic metabolic carcinogenic mechanisms is induction of the mutagenic reactive oxygen species formation by the gut microbiota that affects all vital processes in the colonic epithelial cells (Kumar et al. 2007; Roessner et al. 2008).

The role of gut microbiota in obesity-associated carcinogenesis was also noted. Bäckhed et al. (2004) showed that the amount of fat in the organism of gnotobionts after the transplantation of gut microbiota from conventional laboratory mice was increased by 60% during the following 2 weeks, and they were significantly more resistant to insulin. Moreover, colonization of gnotobionts by microflora possibly associated with obesity resulted in more significant increase of fat in the organism than in the case with colonization by any other microbiota (Bäckhed et al. 2004). These results suggested that gut microflora features are additional factor playing a role in the obesity development.

Chronic inflammation substantially increases the probability of neoplastic transformation of the surrounding cells, inducing mutations and epigenetic alterations through the formation of free radicals, apoptosis inhibition, promotion of lower cell differentiation, stimulation of the proliferation, stroma restructuring, angiogenesis and invasion (Saleh and Trinchieri 2011). Abnormal inflammatory processes in the colon not only interferes with the existence of normal microbiota, but also promotes the colonization of this organ by the pathogens that are absent in normal conditions and cause immune response, exacerbating the inflammation (Sansonetti and Di Santo 2007). In addition, pathogens may induce the inflammation without the preceding alterations of colonic homeostasis, altering the permanent state of physiological inflammation caused by normal microflora to the state of pathological inflammation (Round and Mazmanian 2009). The main mechanisms of such pathological inflammatory reactions are initiated through pattern recognition receptors (PRR), which are possessed as by immune cells as by local epithelial cells (Saleh and Trinchieri 2011) and activate intracellular signaling pathways of proinflammatory cytokine production after the binding of microbial or endogenic ligands (Saleh and Trinchieri 2011). However, the results of various authors investigated the role of these receptors and their pathways in the development of microbiota-induced colorectal cancer are contradictory: Some of them showed the protective role of Toll-like receptor (TLR) signaling pathways (González-Navajas et al. 2010; Salcedo et al. 2010), others demonstrated their procarcinogenic activity (Fukata et al. 2007). With that, *NOD1/CARD4*-knockoutmice (this gene encodes the NOD-like receptor of the same name) were more prone to development of colorectal malignancies after exposure to dextran-sulfatesodium compared to conventional mice under the same conditions (Chen et al. 2008) (nevertheless, tumor frequency decreased after the antibiotic treatment depressing gut microbiota). The initiation of signaling pathways activated by NLRP3 and NLRP6NOD-like receptors may prevent the colorectal cancer since *NLRP3*- and *NLRP6*-knockout mice are characterized by the higher frequency of colorectal malignancies (Allen et al. 2010; Normand et al. 2011).

It seems to be that enhanced PRR activation, induced by significant alterations of qualitative and quantitative microflora composition, may inhibit carcinogenesis, whilst impaired functioning of this receptor group may allow transformed cells avoiding immune system recognition and further death (Saleh and Trinchieri 2011). At the same time, PRR activation may also promote carcinogenesis depending on external factors or features of the receptors itself, creating the proinflammatory microenvironment promoting the tumor growth and chemo resistance development and/or leading to immuno suppression caused by chronic inflammation (Saleh and Trinchieri 2011). In this case, on the contrary, decreased PRR activity minimizes the effects of chronic inflammation and risk of tumor development (Saleh and Trinchieri 2011). Thereby, it is probable that the ideal variant is the "golden mean" representing the balance between low and high activity of PRR-mediated signaling pathways (Saleh and Trinchieri 2011). Obviously, there are complex interactions between the pathways of the inflammation caused by gut microflora that are required further detailed investigation.

In any case, the role of immune mechanisms in colorectal cancer development is proved. For instance, Uronis et al. (2009) showed that *IL-10*-knockout mice (this interleukin suppresses inflammation development) exposed to carcinogenica-zoxyme than were prone to spontaneous colitis development and colorectal carcinoma whilst there were no such alterations in conventional mice exposed to azoxymethan. Interestingly, *IL-10*-knockout gnotobionts had no colitis and tumors after exposure to azoxymethan demonstrating a key role of gut microbiota in development of these pathologies (Uronis et al. 2009). Inaddition, *IL-10*-knockout gnotobionts, colonized only by *Bifidobacterium animalis* had a moderate colitis after certain amount of time in contrast with traditional mice with this bacterium (Moran et al. 2009). It is also known that chronic inflammatory diseases of the gut (ulcerative colitis, Crohn's disease) increase colorectal cancer risk (McConnell and Yang 2009). Taking into account that features of the gut microflora substantially affect the risk of chronic inflammatory bowel diseases (Saleh and Trinchieri 2011), they should also influence the risk of colorectal malignancies.

In addition to metabolic and immune mechanisms, a number of bacteria produce toxins possessing a carcinogenic activity, affecting cell-cell interactions, intracellular signal transduction or induction of mutations and epigenetic alterations (Candela et al. 2011). So, bacterial toxins may function like initiating as well as promoting agents (Candela et al. 2011). It is known that enterotoxigenic *Bacteroides fragilis* produce the fragilysin (Sears 2009) which is matrix metalloproteinase cleaving E-cadherin playing the key role in the regulation of cell-cell interactions and binding to the intracellular α- and β-catenins. Degradation of E-cadherin by fragilysin releases β-catenin which migrates to the nucleus where it activates transcription factors inducing the production of c-myc and cyclin D possessing multiple oncogenic effects (Sears 2009). In addition, this toxin initiates NF-κB- and Stat3-pathway leading to the stimulation of production of proinflammatory cytokines promoting tumor development (Sears 2009). Certain *Escherichia coli* strains are also capable of producing carcinogenic toxins (colibactin may cause double-strand breaksand, hence, oncogenic mutations) (Nougayrède et al. 2006). Cytolethal distending toxin (CDT), produced by as minimum as six species of Gram-negative bacteria (including *E. coli*), possesses similar activity (Lara-Tejero and Galán 2002), lead to an increase in the production of many proinflammatory proteins (TNF-α, NF-κB, IL-6, COX-2) by colonic mucosa, and cause dysplasia in the model of microbial hepatocarcinogenesis (Ge et al. 2008). The similar mechanism is possessed by cycle inhibiting factor (Cif) of many *E. coli* pathogenic strains, which additionally suppresses G2-M-phase transition but does not affecting DNA (Samba-Louaka et al. 2008). In addition, other *E. coli* strains produce cytotoxic necrotizing factor 1 (CNF1), activating proteins of the Rho family which are characterized by GTPase activity and are key for the regulation of many vital processes (proliferation, survival, apoptosis, angiogenesis, motility etc.) (Boquet 2001). It, in turn, initiates a number of intracellular signaling pathways including FAK, Src, COX-2 and NF-κB proteins that are potentially important for neoplastic transformation and mediate a number of intracellular signaling pathways thus resulting in enhanced cell proliferation, apoptosis inhibition, increased proinflammatory cytokine production and other oncogenic alterations (Travaglione et al. 2008). In addition, lipopolysaccharide, which is the main glycolipid component of the Gram-negative bacteria outer membrane, promotes adhesion and invasion of tumor cells via binding to TLR4 that, in turn, leads to NF-κB activation and elevated urokinase plasminogen activator (uPA) production (Killeen et al. 2009). This protein catalyzes the transformation of plasminogen to plasmin degrading the extracellular matrix proteins, enhancing the migration and adhesion of tumor cells (Killeen et al. 2009).

The number of experimental and, particularly, epidemiological studies devoted to this problem significantly increased during the last years. In one of the pioneer investigations, Moore W.E. and Moore L.H. (1995) revealed elevated concentration of *Bacteroides vulgatus*, *Eubacterium* spp., *Ruminococcus* spp., *Streptococcus hansenii*, *Bifidobacterium* spp. And *Faecalibacterium prausnitzii*, and decreased amount of *Lactobacillus S06* and *Eubacterium aerofaciens* in the feces of patients with polyposis. Several studies (Kado et al. 2001; Vannucci et al. 2008; Uronis et al. 2009) showed a decreased colorectal cancer risk in gnotobionts with inherited oncogenic mutations or under exposure to carcinogens compared to the conventional mice. Wu et al. (2009) demonstrated that mice which were long-term artificially colonized by enterotoxigenic *Bacteroides fragilis* developed colitis, and, further, malignant colonic tumors. These results were confirmed by Toprak et al. (2006) who detected enterotoxigenic *Bacteroides fragilis*in 38% of colorectal cancer patients feces (compared to 12% in the control group, OR = 4.16, 95% CI = 1.39–12.43, p = 0.009). Balamurugan et al. (2008) found that colorectal cancer cases are characterized by lower concentration of *Eubacterium rectal* and *Faecalibacterium prausnitzii* and by higher concentration of *Enterococcus faecalis*in feces than in the control group. According to the data of Sobhani et al. (2011), colorectal cancer patients have an elevated concentration of *Bacteroides* and *Prevotella* in their feces. Boleij et al. (2010) revealed the immune response to RpL7/L12 antigen of *Streptococcus bovis* is more enhanced in subjects with polyps and with I/II colorectal cancer stages in comparison with asymptomatic individuals. Scanlan et al. (2008) found that the microbiota of colorectal cancer subjects and patients with polyposis is less stable in time and has greater diversity compared to the microflora of healthy controls. In addition, the diversity of *Clostridium leptum*and *Clostridium coccoides* was also greater in first two groups (Scanlan et al. 2008). In the investigation of Swidsinski et al. (1998), colonicmucosa of colorectal cancer patients contained adhesive and invasive *Escherichia coli* strains in 82% of cases that was significantly greater than in healthy controls. The similar picture was detected when authors compared colorectal adenoma tissues and colonic mucosa of the control group (Swidsinski et al. 1998). Gueimonde et al. (2007) showed that colorectal cancer patients and subjects with inflammatory bowel diseases are characterized by lower counts of *Bifidobacterium* spp., (particularly, *Bifidobacterium longum* and *Bifidobacterium animalis*) in comparison with individuals suffered from diverticulitis. According to the opinion of the authors, it may demonstrate that pathological microbiota alterations can be detected at the earliest stages of multistep carcinogenesis (Gueimonde et al. 2007). Shen et al. (2010) observed that colonicmucosa of patients with colorectal adenoma had higher concentration of *Proteobacteria*, but lower counts of *Bacteroidetes*. In addition, *Dorea* spp., *Shigella* spp. and

Faecalibacterium spp. were detected in adenoma tissues more frequently, but *Bacteroides* spp. and *Coprococcus* spp. were observed less frequently compared to similar normal tissues of the control group (Shen et al. 2010). The quantity and diversity of bacteria in adenoma tissues were higher than in corresponding tissues of the control group (Shen et al. 2010). The results of Wang et al. (2012) who compared bacterial composition of feces of colorectal cancer individuals and the control group (Wang et al. 2012), proved the data of Shen et al. in the relation to *Proteobacteria* and *Bacteroidetes*. In this investigation, colorectal cancer patients were characterized by higher amounts of *Porphyromonas*, *Escherichia/Shigella*, *Klebsiella*, *Enterococcus*, *Streptococcus* and *Peptostreptococcus* and by lower concentration of *Bacteroides*, *Roseburia*, *Lachnospiroceae*, *Alistipes*, *Eubacterium*, and *Parasutterella* (Wang et al. 2012). Interestingly, distinct species of *Bacteroides* significantly differed in concentrations: *Bacteroides fragilis*were more frequent in feces of colorectal cancer patients, *Bacteroides vulgatus*and*Bacteroides uniformis*were more frequent among the control group (Wang et al. 2012). Marchesi et al. (2011) found that the concentration of *Coriobacteridae*, *Roseburia*, *Fusobacterium* and *Faecalibacterium* was higher in colorectal cancer tissues whilst the concentration of *Enterobacteriaceae* (*Citrobacter* spp., *Shigella* spp., *Cronobacter* spp., *Kluyvera* spp., *Serratia*spp. And *Salmonella* spp.) was significantly lower compared to surrounding normal tissues. The authors suggested that since *Coriobacteridae*, *Roseburia*, *Fusobacterium* and *Faecalibacterium* are able to produce short-chain fatty acids, have probiotic features and are commensals, their increased concentration in tumor tissues may be explained by the hypothesis that they colonize them after disease occurrence, supersiding *Enterobacteriaceae* spp., the majority of which possesses certain carcinogenic features (Marchesi et al. 2011). Theoretically, it is possible that microbiota may prevent the tumor from spreading after its occurence (Marchesi et al. 2011). On the other hand, attraction of butyrate-producing bacteria to the tumor may, on the contrary, promote tumor development since butyrate, despite it has anticarcinogenic potential, is simultaneously the energy source for malignant cells and weakens immune response, allowing tumor cells to avoid killing (Marchesi et al. 2011). In one of the recent studies, Castellarin et al. (2012) and Kostic et al. (2012) demonstrated the concentration of *Fusobacterium nucleatum* in colorectal cancer tissues is significantly higher compared to the corresponding normal tissues (79-415-fold), and subjects with higher amount of this bacteria in tumor tissues has higher risk of metastasizing to the regional lymph nodes. Fusobacteria were also visualized within colorectal tumors using FISH. In addition, *Bacteroidetes* and *Firmicutes* phyla were depleted in tumors. The summary and key features of studies indicated above are contained in Table 6.1.

Basing on carcinogenic mechanisms described above, it is obvious that features of qualitative and quantitative composition of gutmicrobiota may significantly affect colorectal cancer risk. Experimental and populational studies carried out in the last years suggested and substantiated the association of the number of its members with risk of this disease. With that, the number of investigations carried out is insufficient to investigate this problem in details. Apart from new cell and animal models of biological carcinogenesis, large epidemiological investigations in various countries for the comparison of interpopulational differences of gut

Table 6.1 The association of the human gut microbiota with colorectal cancer risk in epidemiological studies

Authors, year, population	Sample size and type of biomaterial	Features of gut microbiota in CRC patients
Toprak et al. (2006), Turkish population	73 CRC patients, 59 healthy subjects, feces	>enterotoxigenic *B. fragilis*
Balamurugan et al. (2008), Indian population	20 CRC patients, 17 healthy subjects, feces	>*Enterococcus faecalis* <*Eubacterium rectale* and *Faecalibacterium prausnitzii*
Sobhani et al. (2011), French population	60 CRC patients, 119 healthy subjects, feces	>*Bacteroides/Prevotella*
Scanlan et al. (2008), Irish population	20 CRC patients, 20 healthy subjects, feces	<stability in time >general diversity and diversity among*C. Leptum* and *C. coccoides*
Swidsinski et al. (1998), German population	31 CRC patients, 65 healthy subjects, colonic tissue samples	>adhesive and invasive strains of *Escherichia coli*
Gueimonde et al. (2007), Finnish population	21 CRC patients, 9 patients with diverticulitis, colonic tissue samples	<*Bifidobacterium, B. longum, B. animalis*
Wang et al. (2012), Chinese population	46 CRC patients, 56 healthy subjects, feces	>*Proteobacteria, Porphyromonas, Escherichia/Shigella, Enterococcus, Streptococcus, Peptostreptococcus*and *B. fragilis* <*Bacteroidetes, Bacteroides, Roseburia, Lachnospiroceae, Alistipes, Eubacterium, B. vulgatus* and *B. uniformis*

microbiota influence on colorectal cancer risk are required. More importantly, feces as well as colonic mucosa tissue samples should be a biological material for the analysis because of different composition of the microbiota in these biomaterial types. In addition, because of imperfect study design sometimes it is rather hard to conclude either the member of the gut microbiota plays a role in colorectal cancer etiology or just colonizes the tumor after its occurrence as a result of the local homeostasis disruption. To avoid this problem, not only colorectal cancer patients and healthy controls, but also subjects with inflammatory bowel diseases (ulcerative colitis, Crohn's disease) and individuals with polyposis coli should be included in the study since these pathologies are precancerous for the colorectal malignancies. Ideally, subjects of all four groups should be age-, gender-, and region-matched, but it is feasible to admit just the approximate similarity in age (±2–3 years) because of technical difficulties. The comparison of the gut microbiota features among these four groups will allow analyzing their alterations during the stages of multistep carcinogenesis that has significance for the determination of the etiological role of the gut microflora member and for prevention since in this case it will be possible to calculate colorectal cancer risk and define the risk groups using qualitative and quantitative microbiota alterations.

However, there are multiple interactions between the diet, gut microbiota and colorectal cancer risk, since the composition of food directly affect microflora constitution that, in turn, modulate cancer risk. The investigation of such interactions for the determination of the optimal diet for colorectal cancer prevention, undoubtly, is an intriguing direction in relation to problem considered in this chapter. Moreover, it seems to be that the place of diet in such triad can be superseded by other cancer risk factors (smoking, exposure to professional and environmental carcinogens, low physical activity, hormone alterations, chemotherapeutics, etc.). On the other hand, gutmicrobiota may affect other cancer risk factors (for instance, obesity, production of carcinogenic metabolites, local and systemic inflammation development and synthesis of oncogenic proteins etc.) itself.

One more direction which is a perspective from a theoretical as well as from practical point of view is the feasible role of prebiotics, probiotics and synbiotics in gut microbiota-associated colorectal cancer prevention. It is optimal to combine results of experimental studies on animal models of biological carcinogenesis and epidemiological prospective studies which, however, require a long time of realization in investigating in this direction. This way of prevention may be rather effective and lead to significant decrease of economic losses, incidence, invalidization and mortality from colorectal cancer.

New associations between the gut microbiota members and colorectal cancer risk can be used in the future integrative systems of colorectal risk determination on the basis of these connections, which will be useful to use in various prevention programs. Moreover, since many mechanisms of the gut microbiota-associated biological carcinogenesis are systemic, it is possible to suggest a role of gut microflora in the development of the extraintestinal malignancies. The similar hypothesis should be true for the microbiota of other organs in the case of revealing its systemic carcinogenic mechanisms. The topicality of the problem is underlined by its importance and novelty as from theoretical as from practical point of view. Taking into account the recent tendention to increase the share of biological carcinogenesis-associated cancer cases and continuous progress of the methods of molecular biology, many new associations between human microbiota and cancer risk will be discovered in the future.

References

Allen IC, TeKippe EM, Woodford RM et al (2010) The NLRP3 inflammasome functions as a negative regulator of tumorigenesis during colitis-associated cancer. J Exp Med 207(5): 1045–1056

Andrian E, Grenier D, Rouabhia M (2004) In vitro models of tissue penetration and destruction by *Porphyromonas gingivalis*. Infect Immun 72(8):4689–4698

Azcárate-Peril MA, Sikes M, Bruno-Bárcena JM (2011) The intestinal microbiota, gastrointestinal environment and colorectal cancer: a putative role for probiotics in prevention of colorectal cancer? Am J Physiol Gastrointest Liver Physiol 301(3):G401–G424

Bäckhed F, Ding H, Wang T et al (2004) The gut microbiota as an environmental factor that regulates fat storage. Proc Natl Acad Sci USA 101:15718–15723

Bäckhed F, Ley RE, Sonnenburg JL et al (2005) Host-bacterial mutualism in the human intestine. Science 307(5717):1915–1920

Balamurugan R, Rajendiran E, George S et al (2008) Real-time polymerase chain reaction quantification of specific butyrate-producing bacteria, *Desulfovibrio* and *Enterococcus faecalis* in the feces of patients with colorectal cancer. J Gastroenterol Hepatol 23(8 Pt 1):1298–1303

Bartie KL, Williams DW, Wilson MJ et al (2004) Differential invasion of *Candida albicans* isolates in an in vitro model of oral candidosis. Oral Microbiol Immunol 19(5):293–296

Bernstein H, Bernstein C, Payne CM et al (2005) Bile acids as carcinogens in human gastrointestinal cancers. Mutat Res 589(1):47–65

Bloching M, Reich W, Schubert J et al (2007) The influence of oral hygiene on salivary quality in the Ames Test, as a marker for genotoxic effects. Oral Oncol 43(9):933–939

Boleij A, Roelofs R, Schaeps RM et al (2010) Increased exposure to bacterial antigen RpL7/L12 in early stage colorectal cancer patients. Cancer 116(17):4014–4022

Boquet P (2001) The cytotoxic necrotizing factor 1 (CNF1) from *Escherichia coli*. Toxicon 39(11): 1673–1680

Boyle P, Ferlay J (2005) Cancer incidence and mortality in Europe, 2004. Ann Oncol 16(3):481–488

Byakodi R, Krishnappa R, Keluskar V et al (2011) The microbial flora associated with oral carcinomas. Quintessence Int 42(9):e118–e123

Candela M, Guidotti M, Fabbri A et al (2011) Human intestinal microbiota: cross-talk with the host and its potential role in colorectal cancer. Crit Rev Microbiol 37(1):1–14

Canković M, Bokor-Bratić M (2010) *Candida albicans* infection in patients with oral squamous cell carcinoma. Vojnosanit Pregl 67(9):766–770

Castellarin M, Warren RL, Freeman JD et al (2012) *Fusobacterium nucleatum* infection is prevalent in human colorectal carcinoma. Genome Res 22(2):299–306

Cawson RA (1969) Leukoplakia and oral cancer. Proc R Soc Med 62(6):610–615

Chen GY, Shaw MH, Redondo G et al (2008) The innate immune receptor Nod1 protects the intestine from inflammation-induced tumorigenesis. Cancer Res 68(24):10060–10067

Farrell JJ, Zhang L, Zhou H et al (2012) Variations of oral microbiota are associated with pancreatic diseases including pancreatic cancer. Gut 61(4):582–588

Feng SH, Tsai S, Rodriguez J et al (1999) Mycoplasmal infections prevent apoptosis and induce malignant transformation of interleukin-3-dependent 32D hematopoietic cells. Mol Cell Biol 19(12):7995–8002

Fukata M, Chen A, Vamadevan AS et al (2007) Toll-like receptor-4 promotes the development of colitis-associated colorectal tumors. Gastroenterology 133(6):1869–1881

Ge Z, Schauer DB, Fox JG (2008) In vivo virulence properties of bacterial cytolethal-distending toxin. Cell Microbiol 10(8):1599–1607

Gerlic M, Horowitz J, Horowitz S (2004) *Mycoplasma fermentans* inhibits tumor necrosis factor alpha-induced apoptosis in the human myelomonocytic U937 cell line. Cell Death Differ 11(11):1204–1212

Gill SR, Pop M, Deboy RT et al (2006) Metagenomic analysis of the human distal gut microbiome. Science 312(5778):1355–1359

González-Navajas JM, Fine S, Law J et al (2010) TLR4 signaling in effector CD4+ T cells regulates TCR activation and experimental colitis in mice. J Clin Invest 120(2):570–581

Gueimonde M, Ouwehand A, Huhtinen H et al (2007) Qualitative and quantitative analyses of the bifidobacterial microbiota in the colonic mucosa of patients with colorectal cancer, diverticulitis and inflammatory bowel disease. World J Gastroenterol 13(29):3985–3989

Hirose K, Isogai E, Mizugai H et al (1996) Adhesion of *Porphyromonas gingivalis* fimbriae to human gingival cell line Ca9-22. Oral Microbiol Immunol 11(6):402–406

Homann N, Tillonen J, Meurman JH et al (2000) Increased salivary acetaldehyde levels in heavy drinkers and smokers: a microbiological approach to oral cavity cancer. Carcinogenesis 21(4):663–668

Hooper SJ, Crean SJ, Fardy MJ, Lewis MA, Spratt DA, Wade WG, Wilson MJ (2007) A molecular analysis of the bacteria present within oral squamous cell carcinoma. J Med Microbiol 56(Pt 12):1651–1659

Hooper SJ, Wilson MJ, Crean SJ (2009) Exploring the link between microorganisms and oral cancer: a systematic review of the literature. Head Neck 31(9):1228–1239

Humblot C, Murkovic M, Rigottier-Gois L et al (2007) Beta-glucuronidase in human intestinal microbiota is necessary for the colonic genotoxicity of the food-borne carcinogen 2-amino-3-methylimidazo[4,5-f]quinolone in rats. Carcinogenesis 28(11):2419–2425

Huycke MM, Gaskins HR (2004) Commensal bacteria, redox stress, and colorectal cancer: mechanisms and models. Exp Biol Med (Maywood) 229(7):586–597

Kado S, Uchida K, Funabashi H et al (2001) Intestinal microflora are necessary for development of spontaneous adenocarcinoma of the large intestine in T-cell receptor beta chain and p53 double-knockout mice. Cancer Res 61(6):2395–2398

Katz J, Onate MD, Pauley KM et al (2011) Presence of *Porphyromonas gingivalis* in gingival squamous cell carcinoma. Int J Oral Sci 3(4):209–215

Killeen SD, Wang JH, Andrews EJ et al (2009) Bacterial endotoxin enhances colorectal cancer cell adhesion and invasion through TLR-4 and NF-kappaB-dependent activation of the urokinase plasminogen activator system. Br J Cancer 100(10):1589–1602

Kim DH, Jin YH (2001) Intestinal bacterial beta-glucuronidase activity of patients with colon cancer. Arch Pharm Res 24(6):564–567

Kim YC, Ko Y, Hong SD et al (2010) Presence of *Porphyromonas gingivalis* and plasma cell dominance in gingival tissues with periodontitis. Oral Dis 16(4):375–381

Kostic AD, Gevers D, Pedamallu CS et al (2012) Genomic analysis identifies association of *Fusobacterium* with colorectal carcinoma. Genome Res 22(2):292–298

Krogh P (1990) The role of yeasts in oral cancer by means of endogenous nitrosation. Acta Odontol Scand 48(1):85–88

Kuboniwa M, Hasegawa Y, Mao S et al (2008) *P. gingivalis* accelerates gingival epithelial cell progression through the cell cycle. Microbes Infect 10(2):122–128

Kumar A, Wu H, Collier-Hyams LS et al (2007) Commensal bacteria modulate cullin-dependent signaling via generation of reactive oxygen species. EMBO J 26(21):4457–4466

Kuramitsu HK, Miyakawa H, Qi M et al (2002) Cellular responses to oral pathogens. Ann Periodontol 7(1):90–94

Kurkivuori J, Salaspuro V, Kaihovaara P et al (2007) Acetaldehyde production from ethanol by oral streptococci. Oral Oncol 43(2):181–186

Lara-Tejero M, Galán JE (2002) Cytolethal distending toxin: limited damage as a strategy to modulate cellular functions. Trends Microbiol 10(3):147–152

Ley RE, Hamady M, Lozupone C et al (2008) Evolution of mammals and their gut microbes. Science 320(5883):1647–1651

Lu X, Gao N, Wang C et al (2002) Oral microflora of 42 patients with oral squamous cell carcinoma. Hua Xi Kou Qiang Yi Xue Za Zhi 20(5):356–360

Mager DL, Haffajee AD, Devlin PM et al (2005) The salivary microbiota as a diagnostic indicator of oral cancer: a descriptive, non-randomized study of cancer-free and oral squamous cell carcinoma subjects. J Transl Med 3:27

Mao S, Park Y, Hasegawa Y et al (2007) Intrinsic apoptotic pathways of gingival epithelial cells modulated by *Porphyromonas gingivalis*. Cell Microbiol 9(8):1997–2007

Marchesi JR, Dutilh BE, Hall N et al (2011) Towards the human colorectal cancer microbiome. PLoS One 6(5):e20447

Markowitz VM, Szeto E, Palaniappan K et al (2008) The integrated microbial genomes (IMG) system in 2007: data content and analysis tool extensions. Nucleic Acids Res 36(Database issue):D528–D533

McBain AJ, Macfarlane GT (1998) Ecological and physiological studies on large intestinal bacteria in relation to production of hydrolytic and reductive enzymes involved in formation of genotoxic metabolites. J Med Microbiol 47(5):407–416

McConnell BB, Yang VW (2009) The role of inflammation in the pathogenesis of colorectal cancer. Curr Colorectal Cancer Rep 5(2):69–74

McCullough M, Jaber M, Barrett AW et al (2002) Oral yeast carriage correlates with presence of oral epithelial dysplasia. Oral Oncol 38(4):391–393

Mihara J, Miyazawa Y, Holt SC (1993) Modulation of growth and function of human gingival fibroblasts by fibroblast-activating factor derived from *Porphyromonas gingivalis* W50. Infect Immun 61(2):596–601

Moore WE, Moore LH (1995) Intestinal floras of populations that have a high risk of colon cancer. Appl Environ Microbiol 61(9):3202–3207

Moran JP, Walter J, Tannock GW et al (2009) Bifidobacterium animalis causes extensive duodenitis and mild colonic inflammation in monoassociated interleukin-10-deficient mice. Inflamm Bowel Dis 15(7):1022–1031

Morita E, Narikiyo M, Yano A et al (2003) Different frequencies of *Streptococcus anginosus* infection in oral cancer and esophageal cancer. Cancer Sci 94(6):492–496

Nagy KN, Sonkodi I, Szöke I et al (1998) The microflora associated with human oral carcinomas. Oral Oncol 34(4):304–308

Nakhjiri SF, Park Y, Yilmaz O et al (2001) Inhibition of epithelial cell apoptosis by *Porphyromonas gingivalis*. FEMS Microbiol Lett 200(2):145–149

Narikiyo M, Tanabe C, Yamada Y et al (2004) Frequent and preferential infection of *Treponema denticola*, *Streptococcus mitis*, and *Streptococcus anginosus* in esophageal cancers. Cancer Sci 95(7):569–574

Nieminen MT, Uittamo J, Salaspuro M et al (2009) Acetaldehyde production from ethanol and glucose by non-*Candida albicans* yeasts in vitro. Oral Oncol 45(12):e245–e248

Normand S, Delanoye-Crespin A, Bressenot A et al (2011) Nod-like receptor pyrin domain-containing protein 6 (NLRP6) controls epithelial self-renewal and colorectal carcinogenesis upon injury. Proc Natl Acad Sci USA 108(23):9601–9606

Nougayrède JP, Homburg S, Taieb F et al (2006) *Escherichia coli* induces DNA double-strand breaks in eukaryotic cells. Science 313(5788):848–851

O'Grady JF, Reade PC (1992) *Candida albicans* as a promoter of oral mucosal neoplasia. Carcinogenesis 13(5):783–786

O'Keefe SJ (2008) Nutrition and colonic health: the critical role of the microbiota. Curr Opin Gastroenterol 24(1):51–58

O'Keefe SJ, Ou J, Aufreiter S et al (2009) Products of the colonic microbiota mediate the effects of diet on colon cancer risk. J Nutr 139(11):2044–2048

Prakash S, Rodes L, Coussa-Charley M et al (2011) Gut microbiota: next frontier in understanding human health and development of biotherapeutics. Biol Targets Ther 5:71–86

Putnins EE, Sanaie AR, Wu Q et al (2002) Induction of keratinocyte growth factor 1 expression by lipopolysaccharide is regulated by CD-14 and toll-like receptors 2 and 4. Infect Immun 70(12):6541–6548

Qin J, Li R, Raes J et al (2010) A human gut microbial gene catalogue established by metagenomic sequencing. Nature 464(7285):59–65

Ramasamy S, Singh S, Taniere P et al (2006) Sulfide-detoxifying enzymes in the human colon are decreased in cancer and upregulated in differentiation. Am J Physiol Gastrointest Liver Physiol 291(2):G288–G296

Rautemaa R, Hietanen J, Niissalo S et al (2007) Oral and oesophageal squamous cell carcinoma – a complication or component of autoimmune polyendocrinopathy-candidiasis-ectodermal dystrophy (APECED, APS-I). Oral Oncol 43(6):607–613

Reibel J (2003) Prognosis of oral pre-malignant lesions: significance of clinical, histopathological, and molecular biological characteristics. Crit Rev Oral Biol Med 14(1):47–62

Roessner A, Kuester D, Malfertheiner P et al (2008) Oxidative stress in ulcerative colitis-associated carcinogenesis. Pathol Res Pract 204(7):511–524

Roldán MD, Pérez-Reinado E, Castillo F et al (2008) Reduction of polynitroaromatic compounds: the bacterial nitroreductases. FEMS Microbiol Rev 32(3):474–500

Round JL, Mazmanian SK (2009) The gut microbiota shapes intestinal immune responses during health and disease. Nat Rev Immunol 9(5):313–323

Salcedo R, Worschech A, Cardone M et al (2010) MyD88-mediated signaling prevents development of adenocarcinomas of the colon: role of interleukin 18. J Exp Med 207(8):1625–1636

Saleh M, Trinchieri G (2011) Innate immune mechanisms of colitis and colitis-associated colorectal cancer. Nat Rev Immunol 11(1):9–20

Samba-Louaka A, Nougayrède JP, Watrin C et al (2008) Bacterial cyclomodulin Cif blocks the host cell cycle by stabilizing the cyclin-dependent kinase inhibitors p21 and p27. Cell Microbiol 10(12):2496–2508

Sansonetti PJ, Di Santo JP (2007) Debugging how bacteria manipulate the immune response. Immunity 26(2):149–161

Sasaki M, Ohara-Nemoto Y, Tajika S et al (2001) Antigenic characterisation of a novel *Streptococcus anginosus* antigen that induces nitric oxide synthesis by murine peritoneal exudate cells. J Med Microbiol 50(11):952–958

Sasaki M, Yamaura C, Ohara-Nemoto Y et al (2005) *Streptococcus anginosus* infection in oral cancer and its infection route. Oral Dis 11(3):151–156

Scanlan PD, Shanahan F, Clune Y et al (2008) Culture-independent analysis of the gut microbiota in colorectal cancer and polyposis. Environ Microbiol 10(3):789–798

Sears CL (2009) Enterotoxigenic *Bacteroides fragilis*: a rogue among symbiotes. Clin Microbiol Rev 22(2):349–369

Seitz HK, Simanowski UA, Garzon FT et al (1990) Possible role of acetaldehyde in ethanol-related rectal cocarcinogenesis in the rat. Gastroenterology 98(2):406–413

Shen XJ, Rawls JF, Randall T et al (2010) Molecular characterization of mucosal adherent bacteria and associations with colorectal adenomas. Gut Microbes 1(3):138–147

Sobhani I, Tap J, Roudot-Thoraval F et al (2011) Microbial dysbiosis in colorectal cancer (CRC) patients. PLoS One 6(1):e16393

Sugano N, Yokoyama K, Oshikawa M et al (2003) Detection of *Streptococcus anginosus* and 8-hydroxydeoxyguanosine in saliva. J Oral Sci 45(4):181–184

Swidsinski A, Khilkin M, Kerjaschki D et al (1998) Association between intraepithelial *Escherichia coli* and colorectal cancer. Gastroenterology 115(2):281–286

Takemura A, Matsuda N, Kimura S et al (1998) *Porphyromonas gingivalis* lipopolysaccharide modulates the responsiveness of human periodontal ligament fibroblasts to platelet-derived growth factor. J Periodontal Res 33(7):400–407

Tezal M, Sullivan MA, Hyland A et al (2009) Chronic periodontitis and the incidence of head and neck squamous cell carcinoma. Cancer Epidemiol Biomarkers Prev 18(9):2406–2412

Tillonen J, Homann N, Rautio M et al (1999) Role of yeasts in the salivary acetaldehyde production from ethanol among risk groups for ethanol-associated oral cavity cancer. Alcohol Clin Exp Res 23(8):1409–1415

Tlaskalová-Hogenová H, Stěpánková R, Kozáková H et al (2011) The role of gut microbiota (commensal bacteria) and the mucosal barrier in the pathogenesis of inflammatory and autoimmune diseases and cancer: contribution of germ-free and gnotobiotic animal models of human diseases. Cell Mol Immunol 8(2):110–120

Toprak NU, Yagci A, Gulluoglu BM et al (2006) A possible role of *Bacteroides fragilis* enterotoxin in the aetiology of colorectal cancer. Clin Microbiol Infect 12(8):782–786

Travaglione S, Fabbri A, Fiorentini C (2008) The Rho-activating CNF1 toxin from pathogenic *E. coli*: a risk factor for human cancer development? Infect Agent Cancer 3:4

Uittamo J, Siikala E, Kaihovaara P et al (2009) Chronic candidosis and oral cancer in APECED-patients: production of carcinogenic acetaldehyde from glucose and ethanol by *Candida albicans*. Int J Cancer 124(3):754–756

Uronis JM, Mühlbauer M, Herfarth HH et al (2009) Modulation of the intestinal microbiota alters colitis-associated colorectal cancer susceptibility. PLoS One 4(6):e6026

Vannucci L, Stepankova R, Kozakova H et al (2008) Colorectal carcinogenesis in germ-free and conventionally reared rats: different intestinal environments affect the systemic immunity. Int J Oncol 32(3):609–617

Wang T, Cai G, Qiu Y et al (2012) Structural segregation of gut microbiota between colorectal cancer patients and healthy volunteers. ISME J 6(2):320–329

Williams DW, Bartie KL, Potts AJ et al (2001) Strain persistence of invasive *Candida albicans* in chronic hyperplastic candidosis that underwent malignant change. Gerodontology 18(2):73–78

Wu S, Rhee KJ, Albesiano E et al (2009) A human colonic commensal promotes colon tumorigenesis via activation of T helper type 17T cell responses. Nat Med 15(9):1016–1022

Yilmaz O, Verbeke P, Lamont RJ et al (2006) Intercellular spreading of *Porphyromonas gingivalis* infection in primary gingival epithelial cells. Infect Immun 74(1):703–710

Yumoto H, Nakae H, Yamada M et al (2001) Soluble products from *Eikenella corrodens* stimulate oral epithelial cells to induce inflammatory mediators. Oral Microbiol Immunol 16(5):296–305

Zhang MM, Cheng JQ, Xia L et al (2011) Monitoring intestinal microbiota profile: a promising method for the ultraearly detection of colorectal cancer. Med Hypotheses 76(5):670–672

Zoetendal EG, Rajilic-Stojanovic M, de Vos WM (2008) High-throughput diversity and functionality analysis of the gastrointestinal tract microbiota. Gut 57(11):1605–1615

Chapter 7
Conclusions: Are We There Yet?

Abstract In this chapter, we denote the list of potentially carcinogenic infectious agents which are not included into the IARC list. In addition, we postulate the principles of the study design in the field of biological carcinogenesis.

Keywords Cancer • Infections • Infectious agents • Carcinogens • Biological carcinogenesis • Conclusions • International Agency for Research on Cancer (IARC) • Prevention • Association • List

As was discussed in this book, a lot of associations between infectious agents and cancer risk were discovered through years of cancer research (Table 7.1). It seems to be that the vast majority of new associations of infectious agents and cancer risk would be found among various species of organ microbiota, and some additional ones can also be revealed among protozoa and other bacteria. New tools of molecular biology will also help researchers to achieve this goal. The principles of study designed in the field may be postulated as follows:

- It should investigate metabolic, immune, toxin-related or other, and still unknown, carcinogenic mechanisms (cell lines or animal models)
- Possibly, the interactions between mechanisms of viral and non-viral carcinogenesis, and also between distinct mechanisms of non-viral carcinogenesis can be investigated
- It should explore the influence of organisms on cancer risk in the population (by the comparison of prevalence of the suspicious carcinogenic agent in case and control groups, the number of which, ideally, should comply with the number of stages in the scheme of multistep carcinogenesis)
- It should use the most sensitive and specific markers of infection caused by the microorganism which is under investigation in the study (for instance, PCR or serological tests; in the latter case the diagnostic titer of relevant antibodies should be the most sensitive and specific one)

A.G. Kutikhin et al., *Infectious Agents and Cancer*, DOI 10.1007/978-94-007-5955-8_7,

Table 7.1 Short and extended list of infectious agents that may be a cause of cancer development but are not placed in the IARC roster

Short list		Extended list	
Infectious agent	Cancer type	Infectious agent	Cancer type
Bacteroides fragilis	Colorectal cancer	*Helicobacter bilis*	Gallbladder/bile duct cancer
Streptococcus bovis	Colorectal cancer	*Helicobacter hepaticus*	Gallbladder/bile duct cancer
Mycobacterium tuberculosis	Lung cancer	*Salmonella typhi*	Gallbladder/bile duct cancer
Chlamydia pneumoniae	Lung cancer	*Clostridium paraputrificum*	Colorectal cancer
Chlamydia trachomatis	Cervical cancer	*Clostridium clostridioforme*	Colorectal cancer
Trichomonas vaginalis	Cervical cancer	*Bacteroides* spp. excluding *Bacteroides fragilis*	Colorectal cancer
		Cryptosporidium parvum	Colorectal cancer
		Mycoplasma hominis	Prostate cancer
		Propionibacterium acnes	Prostate cancer
		Neisseria gonorrhoeae	Prostate cancer, cervical cancer, anal cancer, vulvar cancer, bladder cancer
		Treponema pallidum	Prostate cancer, cervical cancer, ovarian cancer
		Ureaplasma urealyticum	Prostate cancer, cervical cancer
		Chlamydia trachomatis	Prostate cancer, ovarian cancer, vulvar cancer, pulmonary MALT lymphoma, esophageal cancer
		Trichomonas vaginalis	Prostate cancer
		Chlamydia pneumoniae	Non-Hodgkin lymphoma
		Chlamydia psittaci	Ocular adnexal MALT lymphoma
		Gardnerella vaginalis	Cervical cancer
		Streptococcus anginosus	Esophageal cancer, oral cancer
		Borrelia burgdorferi	Primary cutaneous B-cell lymphoma
		Bartonella henselae	Endothelial malignancies
		Campylobacter jejuni	Immunoproliferative small intestine disease
		Toxoplasma gondii	Brain cancer

(continued)

Table 7.1 (continued)

Short list		Extended list	
Infectious agent	Cancer type	Infectious agent	Cancer type
		Taenia solium	Brain cancer, haematological malignancies
		Fasciola hepatica	Hepatocellular carcinoma
		Trichostrongylus colubriformis	Epithelial cancer
		Paracoccidioides brasiliensis	Lung cancer
		Porphyromonas gingivalis	Oral cancer
		Candida spp.	Oral cancer

The infectious agents included in the short list may be the most likely associated with cancer risk and progression, and can be the primary targets for investigators. Extended list contains infectious agents which can be noted as the secondary targets for cancer researchers

- In case if it is possible, it should be conducted in various populations (to reach this aim, international consortia may be created)

Discovery of the new connections between various infectious agents and cancer may lead to the boost of efforts to the development of anti-carcinogenic vaccines that, in turn, should result in decrease of cancer incidence around the world. As share of biological carcinogens-related cancer cases still increases through the years, it may be a rather significant and promising way of the war against cancer. At the moment, researchers are only at the first setout of the path of anti-carcinogenic vaccines development (possibly excluding HPV and EBV), but this direction will be undoubtly intensively develop in decades to come.

Index

A.G. Kutikhin et al., *Infectious Agents and Cancer*, DOI 10.1007/978-94-007-5955-8,

MIX
Papier aus verantwortungsvollen Quellen
Paper from responsible sources
FSC® C105338

If you have any concerns about our products,
you can contact us on
ProductSafety@springernature.com

In case Publisher is established outside the EU,
the EU authorized representative is:
Springer Nature Customer Service Center GmbH
Europaplatz 3, 69115 Heidelberg, Germany

Printed by Libri Plureos GmbH
in Hamburg, Germany